THE PICKERING MASTERS

THE WORKS OF CHARLES BABBAGE

Volume 3. *The Analytical Engine and Mechanical Notation*

THE WORKS OF

CHARLES BABBAGE

EDITED BY

MARTIN CAMPBELL-KELLY

VOLUME

3

THE ANALYTICAL ENGINE AND MECHANICAL NOTATION

Routledge

Taylor & Francis Group

LONDON

First published 1989 by Pickering & Chatto (Publishers) Limited

Published 2016 by Routledge
2 Park Square, Milton Park, Abingdon, Oxon OX14 4RN

Routledge is an imprint of the Taylor & Francis Group, an informa business

Babbage, Charles
The works of Charles Babbage
1. Mathematics
I. Title II. Campbell-Kelly, Martin
510 QA37.2
ISBN:978-1-138-76372-2 (hbk)

CONTENTS

CALCULATING ENGINES

SECTION B

The Analytical Engine

SECTION C

Analytical Notation

APPENDIX

SECTION B

The Analytical Engine

I

STATEMENT ADDRESSED TO THE DUKE OF WELLINGTON RESPECTING THE CALCULATING ENGINE[a]

(23 December, 1834)

About 13 years ago, I undertook to superintend the construction of the calculating engine at the wish of the then administration. The grounds on which they took it up were, that it was not in its nature capable of becoming an object of pecuniary profit, that it was of the highest importance to a country, possessing an extensive marine to add to its security by the construction of an engine capable of producing astronomical and nautical tables with unerring precision. I thought, and still think, those were *statesmanlike* reasons.

At all events it could not have been taken up by that administration as a personal favour to one, who was always of liberal though not radical principles. It was taken up for the interest of the country not of the individual; and to suppose the contrary is to maintain the existence of a wrong motive, when the existence of a right one is obvious.

I thought at that time that two, or at the utmost, that three years would suffice for its construction, and with this impression, I made no stipulation for any remuneration for my services, either pecuniary or honorary.

---

[a] This important manuscript is believed to be Babbage's first written allusion to the Analytical Engine. On 23 December, 1834, he addressed this statement to the Prime Minister, the Duke of Wellington, claiming *inter alia* that he had conceived of 'a totally new engine possessing much more extensive powers.' The statement was never published, although it was cited by several writers. Two fair copies of the letter are known to exist, both in the British Library: in Sir Robert Peel's papers (Add. MSS 40611, f.181) and in Babbage's papers (Add. MSS 37188, f.525). The former has been reprinted here, lightly edited for readability.

I continued for above ten years to work amidst the greatest difficulties, at the completion of that engine. I had to communicate with six administrations: I repeatedly and almost continually advanced considerable sums of money from my own private funds, to prevent the delays, which would have arisen, had I waited for the issues from the Treasury.

The best proof that these advances were necessary arises from the fact that since the time at which I resolved to abstain from making any further advances from my private funds, a period of nearly two years, the further construction of the calculating engine has totally ceased. During the previous ten years I have spared no expense in travelling both at home and on the Continent for the purpose of examining all the machinery, whether employed in manufactures or otherwise to which I could get access. The great improvements I have made in the calculating engine, far beyond what I had at first proposed, were one of the consequences of this examination.

The publication of the results of my reflections in the *Economy of Machinery and Manufactures* was another consequence; that work has been translated into most European languages, and has not been entirely unnoticed by the Governments of other countries.

Of the annoyances and difficulties I have experienced during this period, I shall not speak, they are painful to remember, and their relation is useless for the future. Their effect may be seen perhaps in the composition of this paper, and will be more evident in the decision with which I shall act.

Finally, during my superintendance of the engine, I have several times declined situations of considerable pecuniary advantage, in order that my time might be devoted to the completion of that work.

This was more particularly the case in 1824, when a highly advantageous situation was offered to me, which would have produced an income of above £2000 per annum.

About two years ago, I had completed the invention of the engine, the various contrivances were nearly all drawn, a large part of the machinery was executed, and I had caused to be put together a certain portion of it forming a small engine.

This, the first engine that ever calculated mathematical tables, performed its work with perfect precision, and is in my opinion a complete demonstration of the possibility of executing mechanically more than I had promised.

The period had then arrived when I had surmounted all the great difficulties of the question, and it was not unreasonable, even without having made any such stipulation to suppose that these services would have been met with some acknowledgement, at the least that I should have received some testimony of the approbation of my country.

It was not unreasonable to expect, that an administration which prided itself on its desire for the extension of knowledge, should have been happy in an opportunity of distinguishing the first conversion of mental into mechanical processes.

It was certain that any advice addressed to the King in favour of one, whose inventions contributed to give increased security to the sailor, would have found a ready attention.

Some public expression of opinion from the Government became of more importance to me from the circumstances that I was continually attacked by both Tories and Radicals through the press and at public meetings for having received money for my services, a circumstance which would not, I apprehend, have been disgraceful if true, but being utterly false was disgraceful to those who employed it, and did give me some claim to assistance in refuting it, from the country for which I had worked.

I received no such assistance and the engine itself was less noticed by the Government of the greatest naval power in the world, than it would have been in any even of the smallest states in Europe. I mention this as an opinion resulting from my own experience. In countries which I have visited without any introduction, but those afforded by my printed works, and before any portion of the calculating engine was actually completed, I have yet met with attentions, which stand in striking contrast with the neglect I have experienced in my own. I am however bound in justice to your Grace to state that of the seven Prime Ministers who have successively held office during the progress of the calculating engine, you alone seem to have appreciated the nature of the invention, for you alone have personally inspected its progress, and execution.

I shall now proceed to describe the present condition of the calculating engine. A small portion has already been put together and works perfectly.

About twelve thousand parts, in more or less advanced state, together with all the numerous drawings are at present deposited in a fireproof building, which together with workshops and a residence for the superintending engineer have been built for them in East

Street, Manchester Square. These have only been in my custody during the last four months.

There appears to me to be but four ways in which the question relative to this engine can be disposed of:

First, the Government may desire me to continue its construction in the hands of the same person who has hitherto been employed in making it; this is the utmost extent which, in my opinion, the most adverse interpretation of any implied understanding at the commencement of the undertaking could claim from me; and although I should probably accede to it, other considerations render my decision upon that point at present unnecessary. Circumstances have arisen over which neither the Government nor myself have any control, which render that course not only *highly inexpedient*, but I may also say *even practically impossible*; I shall therefore for the present dismiss it.

Secondly, the Government may enquire whether it is not possible to find some other person than the one to whom I have alluded, competent to complete the work: this is a question on which I have for some time bestowed considerable attention. The result of my enquiry is *that it is possible*, that it might be finished in much shorter time and at considerably less expense; but that this course would be attended by great personal sacrifices on my part, and after my experience of the past. I do not feel myself bound to make any further sacrifices. I am ready to point out both the reasons which lead me to the former conclusion, and the means which can be commanded for the completion of the engine, and to consider any proposition which may be made, but I will make none.

A third course is possible. The Government may find some person to supply my place, who may from my designs and drawings undertake to finish the engine. I mention this as a possible case, but I do not imagine the present administration would pursue a course of such doubtful expediency. If however I am mistaken, I am ready at a moment's notice to transfer the engine to any one whom they may appoint.

The last and apparently the only remaining course is to give up the undertaking entirely. I am not unaware of the consequences of such a step; some economist may get up in the House of Commons and enquire why £20,000 has been spent on what he may be pleased to term a useless and an absurd speculation; both parties will be glad to get rid of their portion of the responsibility, and may unite in throwing the blame entirely on me; and that worst of all possible modes of enquiry into *such* a subject, a Committee of the House of Commons, may report that I am wholly wrong. I have experienced the injustice of my countrymen and if need be, I shall not quail before a greater exercise of it.

I come now to a subject of great difficulty, one which I feel that I ought to communicate in order that you may have fairly before you all the circumstances of the case. The difficulty consists in this, that in pointing out remote though possible consequences, I may be accused of using a tone which is quite foreign to my intention, and which any reasonable being must know would be the least likely to influence the decisions of the Duke of Wellington. If anyone should judge thus harshly of me, I must be content to bear it, knowing that it is a tax I must pay to avoid the harsher judgement I should form of myself if I unfairly withheld information from another.

During nearly two years I was deprived entirely of the possession of the whole of the drawings of the calculating engine. About four months since they were delivered up to me, and I immediately began a re-examination and criticism of every part. The result of this, and of my increased knowledge, has been the contrivance of a totally new engine possessing much more extensive powers, and capable of calculations of a nature far more complicated. It is singular that amongst the numerous contrivances which it comprises, no one has yet been adopted, which has been employed in the former one. Much as it exceeds that engine in power, it is not intended to supersede it; on the contrary, it will greatly extend its power and add to its utility.

I have laid aside for a time all other pursuits, and have advanced considerably with the drawings, on which I keep an able draftsman constantly employed. The great mechanical principles and contrivances which regulate its action are already decided upon, and it is my intention to proceed to finish the drawings in such a manner that the whole invention of the new engine shall be complete and susceptible of being executed at any future period.

Whether I shall ever be able to afford to construct such an engine from my own private resources, injured as they have been by the sacrifices I have made in carrying on the former one, is yet uncertain; but it has been suggested to me from several quarters, and has occurred to my own mind, that it is not impossible that the Government of some other country, forming a different estimate, both of the utility of such undertakings, and of the value of the reputation that attends them, might make propositions, which I should be disposed to entertain.

My right to dispose, as I will, of such inventions cannot be contested; it is more sacred in its nature, than any hereditary or acquired property, for they are the absolute creations of my own mind. The consequences of such an arrangement would be, that it would become necessary for me to collect together all that is most excellent in our own workshops; that *methods* and *processes*, which are equally essential to the perfection of machinery, but which are far less easily transmitted from country to country, would be at once brought into successful practice under the eyes and by the hands of foreign workmen; that I should contrive as I have contrived before, new modes of executing work, and that a school of mechanical engineers* might arise, whose influence would give a lasting impulse to the whole of the manufactures of that country; and that the secondary consequence of the acquisition of that calculating engine might become far more valuable than the primary object for which it was sought.

I have now detailed the circumstances attending the present state of the calculating engine; I can make no claim on the Government, because when I undertook it for them, I made no stipulation for any remuneration, either for the invention or for my services in constructing it.

If it has cost the country £20,000, it has been the source of expense to me, which although not of the same extent, is yet considerable compared with my means; and I have made sacrifices to meet that expenditure – it has also prevented me from engaging in occupations by which far more than that sum would have accrued to me.

If it has consumed thirteen years of my life instead of two or three, it will be said that I must suffer the inconvenience arising from my own ignorance. I was ignorant of that which no human being could foresee; I have suffered the consequences.

I feel however that these are not the pleas which it is worthy of a great nation to set up when treating with an individual on such a question. If they are maintained, I ask no favour of that country. New circumstances have now arisen, over which neither party have any control, and a new arrangement must now be made. All I ask is that I

* I have been informed by men who formerly worked at the calculating engine, but who are now scattered about in our manufacturing districts, that they all get higher wages than their fellow workmen in consequence of having worked at that machine.

may have as early a decision of the question respecting the calculating engine, as a due consideration of the circumstances of the case connected with it will admit.

Dorset St, Manchester Square (signed) CHARLES BABBAGE
23 December, 1834

[123/4]

II

UNE LETTRE À M. QUETELET DE M. CH. BABBAGE RELATIVEMENT À LA MACHINE À CALCULER[a]

From *Académie royale des sciences, des lettres et des beaux-arts de Bruxelles, Bulletins*, Vol. 2 (1835), pp. 123–6

Séance générale du 7 et du 8 mai

M. Quetelet lit une lettre qu'il vient de recevoir de M. Ch. Babbage, correspondant de l'Académie, relativement à la machine à calculer, dont différentes circonstances, indépendantes de la volonté de l'auteur, retardent encore l'achèvement. /

Pendant les derniers six mois, dit M. Babbage, j'ai invente une machine nouvelle, d'un pouvoir beaucoup plus grand: j'ai abandonné tout autre sujet de recherches, et j'en fais actuellement les dessins. J'avance rapidement, mais il est probable que je ne la ferai point exécuter ici. Je suis étonné moi-même de la puissance que je suis parvenu à lui donner, et que je n'aurais pas cru possible d'atteindre, il y a un an.

Cette machine est destinée à comprendre *cent* variables (ou nombres susceptibles de changer); et chacun de ces nombres pourra être composé de 25 figures. Elle exécutera alors chacune des opérations suivantes, et les combinaisons qui peuvent en résulter:

$v_1, v_2, \ldots v_{1000}$, étant des nombres quelconques, la machine pourra

Additionner	v_i et v_k;
Soustraire	v_i de v_k;
Multiplier	v_i par v_k;
Diviser	v_i par v_k;
Extraire la racine	v_i de v_k;
Réduire	v_k à zéro.

[a] An English version of this paper appears as the next article in this volume.

Par exemple, si $f(v_1, v_2, \ldots v_n)$, $n < 100$, est une fonction donnée quelconque, qui puisse être formée par addition, multiplication, soustraction, division, extraction de racine, la machine en calculera la valeur numérique. Elle substituera ensuite cette valeur à la place de v_1, ou de toute autre variable, et calculera cette seconde fonction par rapport à v_1.

Il paraît qu'elle peut réduire en tables presque toutes les équations de différences finies. /

Supposez encore que vous ayez, par l'observation, un millier de valeurs de *a, b, c, d*, et que vous désiriez les calculer par la formule

$$p = \sqrt{\frac{a+b}{cd}};$$

la machine sera disposée pour le calcul de la formule, et on y ajustera la première série des valeurs de *a, b, c, d*; puis, elle les calculera, les imprimera et les réduira à zéro; enfin, elle agitera une sonnette pour faire connaître qu'il faut ajuster une nouvelle série de constantes.

Quand il existe une relation entre un nombre quelconque des coefficiens successifs d'une série (pourvu qu'elle puisse être exprimée par addition, soustraction, multiplication, division, puissances ou extraction de racines), la machine en calculera et fera connaître successivement les termes; et on pourra la disposer ensuite de manière à trouver la valeur de la série pour toutes les valeurs de la variable.

Les plus grandes difficultés de l'invention sont déjà surmontées; mais j'aurai besoin de plusieurs mois pour terminer tous les détails et construire les dessins. Cependant les plans sont déjà assez avancés.

M. Babbage fait connaître ensuite qu'il vient de recevoir une lettre de Sir J. Herschel, écrite du Cap de Bonne-Espérance, et qui contient des détails relatifs à la météorologie. Par des comparaisons, déjà faites entre des observations recueillies dans l'Inde et au Cap, Sir J. Herschel est parvenue à des résultats curieux relativement à un mouvement général de l'atmosphère; M. Babbage attend de lui un mémoire sur ce sujet. Sir J. Herschel a porté la Société Philosophique de l'Afrique méridionale à / prendre quatre jours par an (le 21 décembre, de 21 mars, le 21 juin et le 21 septembre), pendant lesquels on fera des observations horaires du baromètre, du thermomètre, etc. Les observations doivent commencer à 6 heures avant midi le 21, en prenant le jour civil et en continuant à compter jusqu'à 6 heures après midi du 22. Si le 21 tombe un dimanche, on ne commencera que le 22. Il est important que l'on choisisse les mêmes jours sur les différens points de la terre.

Les personnes qui prendront part à ces observations, pourront faire parvenir leurs résultats à sir J. Herschel ou à M. Babbage, par l'intermédiaire de l'Académie. /

III

A LETTER TO M. QUETELET FROM CHARLES BABBAGE RESPECTING THE CALCULATING MACHINE[a]

Translated from *Académie royale des sciences, des lettres et des beaux-arts de Bruxelles, Bulletins*, Vol. 2 (1835), pp. 123–6

General meeting, 7 and 8 May

M. Quetelet read a letter which he had just received from M. Ch. Babbage, a correspondent of the Académie, respecting the calculating machine, of which various circumstances, contrary to the author's wishes, are still delaying its completion.

During the last six months I have been contriving another engine of far greater power. I have given up all other pursuits and am making drawings of it and advance rapidly but it is not probable that it will be executed here. I am myself surprised at the powers I have given it which I should not have thought possible twelve month since.

It is intended to be capable of having 100 variables (or numbers susceptible

[a] This article is an English version (not strictly a translation) of 'Une lettre à M. Quetelet de M. Ch. Babbage relativement à la machine à calculer' (1835), which immediately precedes it in this volume. A partial translation of Babbage's letter to Quetelet was first published in Taylor's *Scientific Memoirs* as a preface to the 'Sketch of the Analytical Engine' (1843). A complete translation of the French publication appears in A. W. Van Sinderen's 'Babbage's letter to Quetelet . . .' (*Annals of the History of Computing*, Vol. 5 (1983), pp. 263–7). In preparing this English version for the *Works of Babbage*, use has been made of a letter from Babbage to Quetelet, preserved in the Quetelet Collection in the Bibliothèque Royal de Belgique, Brussels. This letter, which is written in English and dated 27 April, 1835, is believed to be the same that Quetelet read to the general meeting of the Académie, 7–8 May, 1835. The text of this letter has been used, lightly edited for readability, in the version below. Van Sinderen's translation has been used for the French text which did not form part of Babbage's letter. (The editors acknowledge the permission of A. W. Van Sinderen and the editor of the *Annals of the History of Computing* to use their translation.)

of change) placed upon it; each number to consist of 25 figures.

The engine will then perform any of the following operations or any combination of them:

$v_1, v_2, \ldots v_{120}$ being any number, it will

add v_i to v_k;
subtract v_i from v_k;
multiply v_i by v_k;
divide v_i by v_k;
extract the square root of v_k;
reduce v_k to zero.

Hence if $f(v_1, v_2, \ldots v_n)$, $n < 120$, be any given function which can be formed by addition, multiplication, subtraction, division, or extraction of square root the engine will calculate the numerical value of f. It will then substitute this value instead of v_1 or any other variable and calculate the second function with respect to v_1.

It appears that it can tabulate almost any equation of finite differences.

Also suppose you had by observation a thousand values of a, b, c, d, and wish to calculate by the formula

$$p = \sqrt{\frac{a+b}{cd}}.$$

The engine would be set to calculate the formula. It would then have the first set of the values of a, b, c, d put in. It would calculate and print them, then reduce them to zero and ring a bell to inform you that the next set of constants must be put into the engine.

Whenever a relation exists between any number of coefficients of a series (provided it can be expressed by addition, subtraction, multiplication, division, powers, or extraction root) the engine will successively calculate and print those terms and it may then be set to find the value of the series for any values of the variable.

The greater difficulties of contrivance are already overcome but it will take many months to work out all the details and make the drawings; these are however advancing.

M. Babbage then informed us that he has just received a letter from Sir J. Herschel, written from the Cape of Good Hope, which contains some details relating to meteorology. From comparisons, already made between observations gathered in India and at the Cape, Sir J. Herschel arrived at some curious results concerning a general motion of the atmosphere; M. Babbage is waiting for a report from him on this subject. Sir J. Herschel has arranged for the Philosophical Society of South Africa to take four days a year (21 December, 21 March, 21 June, and 21 September) during which they will take hourly readings of the barometer, thermometer, etc. The readings are to start 6 hours before noon on the 21st, taking the whole day, and continuing until 6 hours after noon of the 22nd. If the 21st is a Sunday, they will not begin until the 22nd. It is important that the same days are chosen at different parts of the earth.

The persons who will take part in these observations can send their results to Sir J. Herschel or M. Babbage, using the Académie as intermediary.

IV

ON THE MATHEMATICAL POWERS OF THE CALCULATING ENGINE[a]

(26 December, 1837)

The object of the present volume is to show the degree of assistance which mathematical science is capable of receiving from mechanism. I may possibly in a separate work describe minutely the mechanism I have contrived for that purpose. For the complete understanding of the following pages it will be unnecessary to present to the reader more than a very general outline of the structure of the engine and if he feel indisposed to examine even that, he may pass it over and taking certain mechanical data for granted at once proceed to the mathematical investigation in which he will find that they are all proved to depend on those mechanical data.

The calculating part of the engine may be divided in two portions:

First. The *mill* in which all operations are performed

Secondly. The *store* in which all the numbers are originally placed and to which the numbers computed by the engine are returned.

The plate[b] represents a plan of the engine – those circles placed round

---

[a] 'On the mathematical powers of the calculating engine' is a manuscript in the Buxton MSS collection in the Museum of the History of Science, Oxford. The manuscript was given by Babbage to his friend H. W. Buxton, who quoted it at length in his *Memoirs of the life and labours of the late Charles Babbage* (first published 1987). The complete manuscript was first published in B. Randell, *The origins of digital computers* (Berlin etc., 1973). This version has been lightly edited for readability: the numbering of sections has been made uniform; rules (__) have been used to indicate missing values, and ellipses have been used to indicate unfinished passages in the original. All substantive corrections have been indicated by footnotes.

[b] No plate accompanies the manuscript, but it is likely that Babbage intended to use the general plan, no. 25 dated 6 August, 1840, reproduced as the final plate of the appendix of this volume.

the great central wheel constitute the mill whilst that portion which adjoins the longitudinal part or rack represents the store.

I OF THE MILL

Two axes with figure wheels, **I** the *ingress axis* and "**A** the *egress axis* connect the *mill* with the *store.*

The mill itself consists of:

1. three figure axes
2. three carriage axes
3. ten table figure axes
4. digit counting apparatus
5. selecting apparatus
6. barrels
7. reducing apparatus for barrels
8. operation cards
9. repeating apparatus
10. combinatorial counting apparatus.

1. *Of the figure axes*

The figure axes **A** and '**A** are connected with each other without the intervention of the central wheels so that a number on the figure wheels of one axis may be transferred to those of the other.

These figure wheels are considerably larger than any others in order to allow of sufficient space on their circumference for placing the pinions by which communications are made with other parts of the mill.

By means of some of these pinions a process called *stepping down* and another called *stepping up* may be performed. It consists in shifting each digit of a number one cage lower or one cage higher, which processes are equivalent to the arithmetical operations of dividing or multiplying the number by ten.

Other pinions are fixed on *register axes* **R** and $\mathbf{R}_1$ and convey the two highest figures of the dividend to the *selecting* apparatus. The third figure axis "**A** is placed near the store and constitutes the egress axis. It is adjacent to the digit counting apparatus with which it communicates.

2. *Carriage axes*

These axes **F**, '**F**, "**F** with their peculiar apparatus are employed to execute the carriage of the tens when numbers are added to or subtracted from each other. The carriages **F** and '**F** can be both connected with the figure axis **A** or one of them with the figure axis **A** and the other with '**A** or they may by means of the central wheels be connected with any other part of the mill. The third carriage "**F** is connected both with the mill and the store and may be used with either.

Whenever the number subtracted is greater than that from which it is taken the resulting carriages would if effected, and if the mechanism admitted, produce a carriage in the forty-first cage. This fact is taken advantage of for many purposes – it is one of very great importance and when it happens a *running up* is said to occur. Connected with this part is a lever on which the *running up* warning acts and this lever governs many parts of the engine according as the circumstances demand.

3. *Table figure axes*

These axes are ten in number, nine of them contain the table of the nine multiples of one factor in multiplication and of the divisor in division. The tenth contains the complement of the divisor in the latter operation. They are all connected with the central wheels and the number on each figure wheel can be *stepped up* or *down* upon the other figure wheel of the same cage. The figure which at each stepping goes off from the bottom wheel is transferred to the top wheel.

4. *Of the digit counting apparatus*

This is a mechanism by which the digits of any number brought into the mill may be counted and certain calculations made as to the position of the decimal point in the result of multiplication and division. It is also used to limit the number of figures employed when the engine is making successive approximations, either to the roots of equations or to the values of certain functions. It consists of three distinct systems nearly similar to each other.

5. *Of the selecting apparatus*

When a table of the nine multiples of a multiplicand has been made it becomes necessary in order to effect multiplication to select successively those multiples indicated by the successive digits of the multiplier. This mechanically is not difficult. But when in the process of division it becomes requisite to select that multiple which is next less than the dividend from which it is to be subtracted the mechanical difficulty is of quite a different order and hitherto nothing but the most refined artifices have been found for accomplishing it. This refinement relates however entirely to the *nature* of those contrivances not to the certainty of their action nor to any delicacy of workmanship.

The apparatus consists of a portion of the carrying apparatus for three figure wheels which by the addition of another contrivance renders them available for the purpose of making the selection. This apparatus is placed immediately below the table axes.

6. *Of the barrels*

The barrels are upright cylinders divided into about seventy rings, the circumference of each ring being divided into about eighty parts. A stud may be fixed on any one or more of these portions of each ring. Thus each barrel presents about eighty vertical columns every one of which contains a different combination of fixed studs.

These barrels have two movements:

1. They can advance horizontally by a parallel motion of their axis.
2. They can turn in either direction and to any extent on their axis.

When the barrels advance horizontally these studs act on levers which cause various movements in the mill, the stud belonging to each ring giving a different order.

Amongst these movements or rather these orders for movements the following may be more particularly noticed. The advance of a barrel may order:

1. A number with its sign to be received into the mill from the ingress axis.
2. A number with its sign to be given off from the mill. This number may thus be either altogether obliterated from the mill or it may at the same time be received on the egress wheel or the number may be given

off from the mill to the egress wheel and at the same time be itself retained in the mill.

3. A variable card to be turned.
4. An operation card to be turned.
5. The circular movement of the barrel itself or of any other barrel to another vertical. This always occurs at every step from the beginning to the end of what are called *operations*. The barrels when once ordered by the operation cards from their zero point to any given vertical always direct themselves to be turned to another vertical preparatory to their next advance. This circular motion is however occasionally changed by an action arising from another source.

7. *Of the reducing apparatus*

Behind each barrel is placed a reducing apparatus. It consists of six or eight sectors which can be made to act upon the barrel and give it a rotatory movement so as to make it pass over 1, 2, 3, or any required number of verticals previously to its next advance.

The levers which put these sectors into action are acted upon by:

1. the studs on their own barrel
2. the studs on any other barrel
3. the operating cards
4. the running up levers.

The first and third of these sources of action occur most frequently.

8. *Of the operation cards*

Those who are acquainted with the cards of a Jacquard's loom will readily understand the functions performed by these cards. To those who are unacquainted with that beautiful contrivance it may be necessary to state that the *cards* consist of pieces of thick pasteboard, tin plate, or sheet zinc pierced with a number of holes; these cards being strung together by wire or tape hinges pass over a square prism.

This prism is situated in front of a number of levers placed in rows which govern the reducing apparatus and consequently the barrels. The faces of the prism are perforated so as to present an opening opposite *every* lever.

If the prism alone is made to advance horizontally against these levers then the levers themselves will enter into the holes of the prism and be partly covered by it but they will not be moved out of their places.

Again if a card having as many holes as the prism has, or as there are levers opposite to it, is placed upon the advancing face of the prism no effect can be produced on the levers by this advance of the prism. But if a card having one hole less than the prism is placed on its face then when the prism advances the lever opposite that hole will be pushed and any order given for which that lever was appointed. Suppose after every order the levers to be replaced and let the prism be turned one quarter round; then a new card will be presented to the levers and if one or more holes of this second card are stopped up a different order will be transmitted through the levers to the reducing apparatus and thence to the barrels.

Thus by arranging a string of cards with properly prepared holes any series of orders however arbitrary and however extensive may be given through the intervention of these levers.

The number of the levers acted upon by the operation card is small; they respectively direct the barrels to *commence* the following operations:

1. the *addition* of two numbers
2. the *subtraction* of one number from another
3. the *multiplication* of two numbers
4. the same *multiplication limited* to a given number of the first figures
5. the *division* of one number by another
6. the same *division limited* to a given number of figures in the quotient.

The levers numbered 4 and 6 are rarely used; the extraction of roots being the only case in which they are required.

These cards are called into action by orders from the barrels. What they shall order when acting depends on the nature of each individual card. What repetitions they shall be subject to depends on the orders communicated to them from the combinatorial cards and their counting apparatus. Many calculations are much simplified by having two sets of variable cards.

9. *Of the combinatorial cards*

One or more peculiar cards may be inserted amongst the operation cards of certain formulae. They are called *combinatorial cards*. The object of these cards is:

To govern the *repeating apparatus* of the operation and of the variable cards and thus to direct at certain intervals the return of those cards to given places and to direct the number and nature of the repetitions which are to be made by those cards.

Whenever combinatorial cards are used other cards called index cards must occur amongst those of the formulae. The use of these cards is to compute the numbers which are to serve successively for the indices of the combinatorial cards. At what time the combinatorial cards shall act depends on the number of repetitions the last of those cards appointed. What orders each combinatorial card shall give depends on the nature of each individual card.

II OF THE STORE

The store may be considered as the place of deposit in which the numbers and quantities given by the conditions of the question are originally placed, in which all the intermediate results are provisionally preserved and in which at the termination all the required results are found.

Various parts may be added to the store according to the purposes required. Some of them might perhaps with more convenience constitute distinct machines.

The store then may contain:

1. figure axes
2. computing apparatus
3. number cards
4. card punching apparatus
5. printing apparatus
6. copper punching apparatus
7. curve drawing apparatus
8. variable cards.

The figure axes and the variable cards alone are absolutely necessary for the mathematical enquiries in the present work and for the sake of simplicity the others will be only occasionally referred to.

1. *Of the figure axes*

A number of axes each having forty figure wheels placed in different cages one above another are connected with the rack of the store.

These figure wheels are each numbered from 0 to 9; they may be turned by hand so that any digit may stand opposite a fixed index. Thus any number of not more than forty places of figures may be put upon the figure wheels of each axis.

Above the fortieth cage is another cage containing a wheel similar to a figure wheel and also having its circumference divided into ten parts. These parts have the signs (+) plus and (−) minus alternately engraved upon them. Above this wheel is a fixed character to distinguish each particular axis or rather the variable number which may be found upon

	Variables of the Engine	V_1	V_2	V_3	V_4	V_5	V_6
Signs	Cage 41	+	−	+	−	+	+
Numerical coefficients	Cage 40	0	0	0	0	0	0
	Cage 39	0	0	0	0	0	0
	⋮	⋮	⋮	⋮	⋮	⋮	⋮
	Cage 5 Tens of thousands	0	0	0	2	0	0
	Cage 4 Thousands	0	1	4	3	0	0
	Cage 3 Hundreds	0	7	9	4	0	2
	Cage 2 Tens	0	5	7	1	3	4
	Cage 1 Units	0	8	1	0	6	7
Variables of the question			a	x	x^3	$\sin\theta$	$\sqrt{a^2 + x^2}$

its wheels. These fixed marks are V_1, V_2, V_3, ... as far as the number of quantities which can be contained in the store.

Below the lowest or units figure wheel a small square frame appears in which may be inserted a card to be changed according to the nature of the calculation directed. On this card is written that particular variable or constant of the formula to be computed whose numerical coefficient and sign are expressed on the wheels above it.

The annexed representation will perhaps convey a clearer idea of this part of the engine.

The first line of quantities V_1, V_2, ... are never altered; they merely indicate the particular sets of figure wheels to which they are attached.

The next line contains the *signs* of the quantities which are themselves expressed by writing them upon pieces of card and placing them in the squares at the bottom.

The intermediate forty cages contain the numerical coefficients.

The first variable or V_1 is in the present figure equal to zero and previously to the setting of the engine to any problem; all the other variables V_2, V_3, V_4, ... have the same symbols beneath them.

The second variable V_2 has beneath it the negative sign, the number 1758 and the algebraic quantity a.

The symbols placed on the engine are in the annexed figure:

$$\begin{array}{ll} V_1 = 0, & \\ V_2 = -1758, & a \\ V_3 = +4971, & x \\ V_4 = -23410, & x^3 \\ V_5 = +36, & \sin\theta \\ V_6 = +247, & \sqrt{a^2 + x^2} \end{array}$$

In this manner any function however complicated it may be, if it is considered as a whole, may be placed in the store with its proper sign and its numerical coefficient; the function itself being merely written on a card and placed in the square below its coefficient. The number of variables which can be contained within the store will depend on the length of the rack and number of figure axes which can be placed round it, and although a large number of variables might with perfect safety be employed yet there is obviously a practical limit arising from the weight of the rack to be moved.

One hundred variables would not give an inconveniently large rack but still the calculations of such an engine would be limited. This limitation can be entirely removed by another set of cards called number cards which will presently be described.

If any of the coefficients contain decimals or if the result is required with any number of decimals then all the coefficients must be considered as having the same number of decimals. If an imaginary line is drawn between any two cages, the third and fourth for example, then all below it may be considered as decimals. In order to convey to the engine this information there exists a wheel with the numbers from 1 to 40 engraved on its edge; this wheel being set at any number the engine will treat all the numbers put into the store as having that number of decimals.

2. *Of the computing apparatus*

One of the figure axes **K** has its figure wheel connected with the rack differently from the others. It is also connected with the carrying axis ″**F** belonging to the mill from which however it may at times be disconnected.

The object of this is to enable the figure wheel on the rack to make certain simple calculations without the necessity of sending the numbers into the mill. During the process of one of the great operations in the mill a series of numbers may be computed in the store by the method of differences and thus a considerable saving of time effected.

3. *Of the number cards*

The number cards have been introduced for the purpose of rendering the calculations absolutely unlimited by the too great number of variables and constants necessary for the solution of any problem.

The number cards are pierced with certain holes and stand opposite levers connected with a set of figure wheels placed on the number axis which can be made at intervals to communicate with the rack. When these number cards are advanced they push in those levers opposite to which there are no holes on the cards and thus transfer that number together with its sign which the holes on the card represent to the figure wheels of the axis opposite to which the cards are placed. The number and its sign thus put upon these figure wheels may be immediately transferred to another part of the store, and the string of number cards being turned the next card conveys its number to the store in the same way.

These number cards are for some purposes more convenient than figure wheels because the numbers upon a figure card are not obliterated by the act of giving them off. For by turning the string of figure cards back to any given one the number upon that card can be replaced in the store as frequently as may be required and at any periods of time which the calculation may demand. On the other hand the numbers placed upon figure wheels are always obliterated in the act of giving them off. If it is necessary to retain on the store any number which is to be given off to the mill then it also must be given off through the rack to another store axis on whose figure wheels it must remain until at a second operation it is reconveyed by rack back to its original place.

4. *Card punching apparatus*

One mode of rendering permanent the results of any calculations made by the engine will obviously be by making it punch on cards certain holes similar to those just described as existing on the number cards. This plan will also enable us to use any intermediate computation which may be necessary in advancing towards the final results, for the cards so made may from time to time be removed from the punching apparatus and attached to the number cards. Other advantages will be observed when the subject of mathematical tables comes under consideration.

5. *Printing apparatus*

It is desirable even when many copies of a calculation made by the engine are not wanted that the results should be themselves printed by the machine in order to ensure the absence of error from copying its answers.

A set of thin circular rings having metal types of the digits fixed at equal distances on their edges and themselves governed by the calculating wheels on which the result is placed are to be pressed down at intervals on a sheet of paper covered by another sheet of carbonized paper. This paper is fixed in a platform having proper motions for placing the printed results in right order.

Thus a single and correct copy may be produced although from the nature of the process the execution of the printing would not be of the highest order.

If however many printed copies are required then it is intended that the type so arranged shall be made to impress their characters on a soft substance from which mould a stereotype plate may be cast.

6. *Copperplate punching apparatus*

If it should be deemed necessary to print tables or calculations upon copperplate an apparatus has been contrived for that purpose. This process is necessarily slower in its operation then the former modes of rendering the calculated results permanent. It has however the advantage of possessing greater clearness although the additional cost of taking off impression may in some instances be objectionable.

Since, however, the invention of the number cards these modes of printing or engraving have ceased to become essential parts of a calculating engine. The absolute certainty of every printed result can now be obtained although the printing mechanism be totally detached from the calculating portion of the engine, an improvement which it was impossible to make until that point of the enquiry was attained.

The cards on which the results are punched may themselves be placed in a distinct machine and from the holes formed in them the new machine may either engrave or print them as it may have been prepared to operate.

7. *Curve drawing apparatus*

The discovery of laws from the examination of a multitude of tabulated and reduced observations is greatly assisted by the representation of such tables in the form of curves.

As one of the employments of a calculating engine would be to reduce collections of facts by some common formula, I thought that at the time it impressed the computed results it would be desirable that it should mark the point of a corresponding curve upon paper or copper if preferred. The three or four first figures of the table will be expressed by the curve. The contrivances for this purpose are not difficult and their employment does not lengthen the time of the calculation.

8. *Variable cards*

The variable cards are appointed for the government of the various parts which constitute the store. Like all the other cards they act by pushing forward certain levers placed in front of them. These levers cause the motions of the several parts of the store which have been described.

It is necessary in the present work to consider only: the figure axes; the number, variable, combinatorial, and operation cards; and the card punching apparatus.[a]

With respect to these, the principle functions of the variable cards will be to direct:

1. A number and its sign to be given off from any store axis to the ingress axis.
2. A number and its sign to be received upon any store axis.
3. Any number and its sign to be given off from the number cards to the wheels on the number axis.
4. Any number and its sign to be given off to the card punching apparatus and a corresponding card to be punched.

The number of levers necessary for these purposes is not so large as might at first appear, consequently the cards need not approach an inconvenient magnitude. For example fourteen levers and their equivalent fourteen holes will be all that is required in the third of the above divisions for eight thousand variables.

If the other appendages to the store which have been already described should be thought necessary a small number of additional levers must be added.

[a] This paragraph has been reworded to incorporate an addendum to the manuscript.

III OF OPERATIONS

The operations which can be performed in the mill are the following:

addition
subtraction
multiplication
division
extraction of roots.

These five processes are all which it will be necessary to consider in the present volume. Others might be added if required and the additional mechanism would be small, new barrels with studs differently arranged would be nearly the whole change.

If these operations are merely viewed in an arithmetical light the engine would greatly assist us in the numerical calculations to which analysis leads, but it possesses the power of treating the signs of the quantities on which it operates according to the rules of algebra and thus its use is greatly extended.

1. *Addition*

There are three principles on which mechanical modes of adding numbers together may be constructed.

An axis passing through a wheel whose circumference is divided into ten parts may be conceived at certain periods to make a revolution independently of the wheel. This axis therefore will, whenever it moves, have an angular motion of one circumference which is equal to ten divisions on the wheel.

Now if a stud projects from the upper surface of the wheel and if a fixed arm project from the axis in the same plane, then whenever the axis revolves its arm will at some period in its revolution come into contact with the stud on the wheel and will carry it forward until the axis stops.

If the wheel had been so placed previously to the movement of the axis that the stud upon the wheel stood *one* division in advance of the arm then the wheel would have been driven over nine divisions; if the wheel had been so placed at the commencement that the stud had been *six* divisions in advance of the arm then it would have been moved over four divisions. In the first case the number *nine* would have been added to the wheel and in the latter case the number *four*. This addition might be communicated by a pinion to any other wheel.

Thus by rightly placing the first wheel, any number from one to nine may be added to a second wheel. When such mechanism is employed the axis moves uniformly round and picks up the wheel at a time dependent on the digit it is to add. Also the motions of the wheel and of the axis terminate together.

Another method of adding might be imagined in which the axis should be bolted to the wheel at the commencement of its motion and carry it along for a certain part of its motion only, some subsidiary mechanism unbolting it when the wheel had reached a certain division.

In this case whatever be the number added the motion of the wheel commences with that of the axis but ends at different periods according to the nature of the number added. The combination of these two principles produces a third mode of adding in which neither the beginning nor the end of the additions occur at fixed points.

The three principles then by which addition may be executed mechanically may be thus stated:

1. By the movement of a wheel which always ends its motion at a definite time independently of the particular number to be added but which commences its motion at a time determined by that number.

2. By the movement of a wheel which always commences its motion at a definite time but ends it at a period dependent on the particular number added.

3. By a wheel both commencing and terminating its movement at periods dependent on the number added.

The first of these principles is adopted in the present engine. The second exists in the difference machine. The third presents greater difficulties in the execution without equivalent advantages.

In all these plans the number placed on the first wheel and which is to be added to some other number is itself obliterated by the operation. They might therefore with propriety be called apparatus for reducing figure wheels to zero. Another class of contrivances has been drawn in which the number added is itself retained so that the same number may be repeatedly added. The zero apparatus may generally be converted into an adding and retaining apparatus by the addition of two other wheels.

The process of addition requires for its completion that an apparatus should be attached to the wheels on which the sum is placed for the purpose of carrying any tens that may occur to the next highest digit.

This mechanism may be constructed upon three different principles. It may be: successive in its operation; it may postpone its operations; it may anticipate operations.

Carriage may be successive in its operations. Those of the first class are the most simple but the time necessary for the performance is an effective impediment to their employment in engines like that we are considering. Let us suppose unity to be added to 999999 then

Add together	1
	00999999
Time of 1st carriage	00999990
	1
Time of 2nd carriage	00999900
	1
Time of 3rd carriage	00999000
	1
Time of 4th carriage	00990000
	1
Time of 5th carriage	00900000
	1
Time of 6th carriage	00000000
	1
	01000000

The time necessary for the mere addition and leaving a notice that the first figure in the right has passed a ten (which is indicated by putting the unity below the tens figure) must be at least equal to that which is required for the figure wheel to pass over nine divisions. Now the time for adding the first ten which is carried and giving notice of the second will be equal to one division and so on, so that at the end of the sixth carriage a time equal to fifteen divisions will have been passed over and it is to be observed that another division must be allowed because the machine cannot know that there is not another carriage due to the seventh figure.

Thus in an engine having forty figures we should have for every addition and its carriage

for addition	9
for 39 possible carriages	39
	—
	48

So that in fact the time consumed in making the carriage of the tens would be more than four times as much as that required for addition. Modes have been contrived for shortening this time but it is evident there is a limit beyond which it cannot be reduced.

Carriage may postpone its operations. When many additions are to be made to the same quantity, as in multiplication, time might be saved by reserving the notices of carriage and executing them altogether.

Carriage may anticipate its operations. If the mechanism which carries could be made to foresee* that its own carriage of a ten to the digit above when that digit happens to be a nine would at the next step give notice of a new carriage then a contrivance might be made by which, acting on that knowledge, it should effect both carriages at once. Thus, after the addition of the numbers below

Time of addition	123456
	346601
Time of 1st carriage	469057
	1
Time of 2nd carriage	460057
	1
	470057

* In substituting mechanism for the performance of operations hitherto executed by intellectual labour it is continually necessary to speak of contrivances by which certain alterations in parts of the machine enable it to execute or refrain from executing particular functions. The analogy between these acts and the operations of mind almost forced upon me the figurative employment of the same terms. They were found at once convenient and expressive and I prefer continuing their use rather than substituting lengthened circumlocutions.

For instance, the expression 'the engine *knows*, etc.' means that one out of many possible results of its calculations has happened and that a certain change in its arrangement has taken place by which it is compelled to carry on the next computation in a certain appointed way.

If at the time of the first carriage the engine knew the next number to which it was about to carry was a nine and that consequently it would *afterwards* become necessary to give notice of and to carry another ten then it might be taught to execute both these carriages at the same time and consequently to anticipate the time of the second carriage.

This object is of such importance that it is worthy of any labour to obtain a plan at once simple and effective. I have contrived several; that which is adopted in the present engine is perfectly secure and considering the very complicated nature of the conditions it is not very deficient in simplicity. This extreme complication of the conditions will perhaps be more felt by considering the following cases[a]

```
Time of addition      006012345678
                      001987655022
                      ____________
                      007999990690
Time of carriage             1  1
                      ____________
                      008000000700

Time of addition      0004365780064128976438052
                      0005634238990771023562048
                      ________________________
                      0009999918954899999990090
                             1  1        1  1
                      ________________________
                      0010000019054900000000100
```

The complexity arises from the circumstance that the result of any carriage *may* effect the value of *every* figure on its left hand and that the effect of *any* carriage may terminate at any digit and a *new* carriage may commence its effect at the succeeding one; and this like its predecessor may or may not be the precursor of many others.

2. *Of the algebraic signs as used in addition*

It has been already stated that above every set of forty figure wheels there is in the 41st cage a sign wheel and also that this sign wheel is divided like the figure wheels into a number of parts which is equal to

[a] The diagram has been corrected.

some multiple of ten, the sign minus being engraved on the odd, the sign plus on the even divisions.

Now when these wheels all stand at zero the sign wheel being also zero will present the positive sign.

When a positive number is added its sign being fixed on some even number the sign wheel of the ingress axis will be advanced by an even number of digits and must consequently present some other even number on which will appear the sign *plus*.

The ingress wheels of the mill which receive any number receive also its sign, but as the number on the ingress wheels was zero and the sign plus, the addition merely causes the ingress wheels to stand at the number entering and the sign to remain as it was before, namely at plus. For the 41st wheel stood at + or at an even number, the addition of + or any other even number will produce an even number on this wheel and consequently it will still present the sign plus. If on the other hand the number entering has the negative sign then its sign wheel stands at an odd division and as an odd number added to an even one produces an odd number the sign of the number in the ingress wheels will be negative.

If therefore when the first of two quantities, which it is proposed to add together, is placed upon the ingress wheels the sign wheel stands at an even number, *no change* is made on the barrels; but if, when the sign of that quantity is negative or the sign wheel of the ingress axis stands at an odd division, a *certain change* is made, then the mill may be said to *know* that a negative quantity has entered and the barrels may give the orders necessary in consequence.

If two numbers P and Q are to be added together the result will be

$$+P+Q$$

where the two signs + + may be called the algebraic signs; but if both P and Q are the result of certain preliminary computations made in the engine it may not be known whether the numbers themselves denoted by P and Q are positive or negative. In this case we must distinguish between the known algebraic signs and the possible or contingent signs resulting from the computation – these latter may be called the *accidental signs*. The above formulae may therefore be written thus

$$+(\pm P)+(\pm Q)$$

The process of addition or subtraction is thus executed:

An operation card advances and orders the variable cards to advance;

in the next turn a variable card containing the order for the sign plus or minus advances and places the sign on the sign wheels of ″**F** whilst at the same time it had ordered the number P to be given off from the store to **I**. The sign on the card may be called the algebraic sign. When the number comes in from the store it will bring with it its own proper sign which will depend on the previous operations by which it had been formed, and it will be convenient to call this latter its accidental sign.

This accidental sign will be added to the algebraic sign upon ″**F** and thus the two signs will be united and we shall have $\pm(\pm P)$ reduced to $\pm P$, any combination of the above signs being used.

If algebraic and accidental signs are of the same kind P is added to the figure wheels on **A** which have a positive sign on them. If the algebraic and accidental signs are different P is subtracted from the wheel on **A** and in consequence of a running up the negative sign is put upon the sign wheel of **A**.

The same process takes place with respect to $\pm(\pm Q)$ which after the combination of its two signs is transferred either additively or subtractively to the same wheels of **A**. In this second case a running up may or may not take place; if it does the previous sign is reversed.

The sum of two numbers P and Q as modified by their respective signs will now be found on the figure wheels of **A** either in the form of a number or its complement.

If it is number it and its sign will be transferred to ″**A**.

If it is complement by being subtracted from ″**A** it will appear as number and a negative sign will appear on the sign wheel.

This being premised the process of addition may be thus stated:

At the first turn of the handle an operation card is turned; it advances and acting on the barrels through the reducing apparatus orders them to move to that vertical which commences addition. It also orders a variable card to advance. At the next turn of the handle the barrels advance and direct the ingress figure wheels to receive from the store any number and its sign which the variable cards may have ordered to be given off. The barrels in advancing order themselves to turn circularly after their advance so as to present the next vertical necessary for performing addition. If the *accidental sign* of the number on the ingress wheels is positive this arrangement continues, but if that sign is negative it effects a change in the reducing apparatus and before the next turn the barrels have been moved to a different vertical which gives the orders necessary from the fact of the accidental sign of the number brought in being negative.

At the third turn the barrels advance and order the figure wheels on **A** and on the two carriages **F**, '**F** to receive the number and its sign, both duly modified, from the ingress figure wheels whilst at the same time the variable card which was turned in the last turn of the handle orders a new quantity Q and its accidental sign out of the store onto the other figure wheels of the ingress axis. But this second advance of the barrels has ordered a circular motion for themselves by which they would turn to the next vertical for addition whilst the *accidental sign* of the quantity Q introduced on the ingress wheels again modifies this order if it happens to be negative.

...

3. *Subtraction*

After the explanation of the process of addition with the accidental sign of the quantities added together it will not be difficult to understand that the operation of subtraction is carried on in a similar manner. As in addition the subtraction of a greater number from a less (in that case arising from the accidental signs) causes a running up which gives notice that the resulting quantity is the complement of a negative number; so in subtraction where the algebraic sign produces the same result a similar effect follows. The notations which follow will explain more clearly the several stages of the operation as well as the parts taken in it by the several portions of the mill.

4. *Of the time required for the operations of addition and subtraction*

Any two numbers P and Q whose sum does not exceed forty places of figures may be placed in the engine and whatever be the nature of their accidental signs they may be brought out of the store placed in the mill and the operation of addition or subtraction performed as indicated by the formula

$$\pm(\pm P) \pm (\pm Q)$$

and the result may be returned into the store in six turns of the handle. The first of these turns might take place as the final turn of any previous operation so that the time really required is six turns of the handle. If another quantity $(\pm R)$ is to be added or subtracted it will require another turn and generally if n quantities with their accidental signs are to be added and subtracted the time required for the operations will be $n + 4$ turns of the first mover.

The time which the edge of the figure wheel occupies to pass from one figure to the next figure is considered as the unit of time and the space between two adjacent figures is the unit of space. The cases in which that velocity is exceeded are very rare in any part of the machine and only occur when the moving part is carried on to a fixed stop. Wishing to under- rather than overrate the speed of its calculations I have assumed that the velocity of the circumference of the figure wheel when moving will be ten feet or one hundred and twenty inches per minute. A single turn of the first mover will according to circumstances be performed either in fifteen or in twenty of such units. In the process of addition cycles of twenty units only are employed and as each unit requires ·15708 seconds the addition of two numbers only will require __ seconds and the addition of n quantities will cost 21·99 or nearly 22 seconds.

5. *Multiplication*

The process of multiplication is thus performed by the engine:

The two factors P and Q with their respective accidental signs being placed on the two sets of figure wheels No. 1 and No. 2 of the ingress axis, an operation card is turned. This directs the barrels through the reducing apparatus to move circularly to the vertical which commences multiplication.

At the next turn the barrels advance and order the reception of one of the factors Q. They also direct themselves to move on their axes to the second vertical belonging to multiplication.

At the third turn the barrels advance and order the reception of the other factor P from the second set of ingress wheels. At this turn the factor P is subtracted from the factor Q which had at the preceeding turn been placed upon the figure wheels of 'F. The result of this subtraction determines which of the two factors is largest and consequently which factor is to be tabulated.

Much time is saved by this decision for supposing one factor to contain only four places of figures and the other to contain thirty five figures. Then if a table of the first nine multiples of the first factor were made its multiplication by the other factor would require thirty-five-additions whilst if the larger factor had been tabulated only four additions would have been necessary. The mechanical mode by which this knowledge is conveyed to the barrels is thus. The barrels, after ordering the subtraction of P from Q in the axis 'F, direct themselves to move on to another vertical; if P is less than Q no *running up* takes place and the order thus given by the barrels is obeyed; but if P is larger than Q then a *running up* takes place and the order given by the barrels to move to a certain vertical is enlarged and they are directed by the running up lever to move on to a different vertical. This new vertical directs the tabulation of the second factor whilst the former vertical appointed that of the first factor.

At the next turn the number which the mill has decided on tabulating is added to each of the nine table axes and to the carriage **F**.

At the next turn the same number is added to eight of the table axes and the carriage axis, then to seven, and so on until the nine multiples have all been formed in the nine table axes.

During these turns the number of the digits in the multiplier has been ascertained and placed in the counting apparatus No. 1.

The multiplier having been prepared is now stepped down and acts upon the *selecting* apparatus. Whatever digit exists in the lowest figure of the multiplier, by means of the selecting apparatus the corresponding multiple is added over to the product axis '**A**. During this addition each multiple is stepped up one cage higher in its own figure wheels; also the multiplier being stepped down its next figure is conveyed to the selecting apparatus preparatory to the next addition to the product.

The product is thus formed beginning with its lowest figure on the figure wheels of the axis **A**, but in the progress of stepping up the highest figure on the table axes may at length reach the top or 40th cage of their axes. At the next stepping the highest figure would of course be stepped off the machine and lost but this is prevented by a communication with the lowest figure wheel on the same axis to which the number stepped off from the highest is transferred. These figures which are so transferred to the lower cages are not added to the same axis but are by a particular mechanism added over to a set of figure wheels on the axis **A** and the carriage **F** on which the head of the products is formed.

If a number of twenty figures only is multiplied by another factor containing an equal number of digits, as the product will not exceed forty digits it will be contained on the first axis. But if both factors contain forty digits then the head or first forty digits of the product will be formed on the axis **A** and the tail or second forty digits of the products will be formed upon '**A**.

The signs which entered with the original factors, having been added together on the sign wheels, will give the resulting sign of the product. For if they were both positive then each sign being equivalent to an even number, the sum of the two signs would give an even number or its equivalent the positive sign. If on the other hand both signs were negative then a negative sign being equivalent to an odd number, the sum of two odd produces an even number which also is equivalent to the positive sign which the product ought in this case to possess. Lastly, if the sign of one factor be positive and that of the other negative, then the sum of an even and of an odd number being odd, the negative sign results.

The termination of multiplication arises from the action of the counting apparatus which at a certain time directs the barrels to order the product thus obtained to be stepped down so that the decimal point may be in its proper place, and then to turn to another vertical, the consequence of which is the transfer of the product with its proper sign to the wheels on the egress axis from whence they may be removed by variable cards and the barrels to any axes in the store.

The great difficulties in constructing the calculating engine have in general arisen from certain conditions I had laid down as desirable for the perfect security and certainty of its action rather than from the difficulty of inventing mechanical contrivances for performing the functions of its several parts. There are few of its actions which have not been contrived and in many instances drawn according to several different principles. The only exceptions I should be inclined to make as to this relative difficulty are perhaps the mode of carrying the tens by anticipation and amongst the operations the mode of dividing one number by another.

The following principles were proposed:

No movement should depend upon the action of a spring.

This condition placed great difficulties in the way of many contrivances and the modes of obviating them are very various. It is not however proposed to exclude *all* springs, for those which merely retain a lever or a wheel that has been already brought to its exact place by other means are unexceptionable. In one sense they relieve the machine from the continual drag to which stiff friction if substituted would expose it, and on the other they save a multitude of movable axes which would otherwise become necessary and they also enable wheels to remain retained until certain indefinite actions take place.

Another principle is that:

Every movement shall be of such a kind that the engine shall either break itself or stop itself or execute the intended motion.

A third principle is:

That the several operations shall be executed in the shortest possible time.

This last principle has caused greater difficulties than either of the former – in fact the whole history of the invention has been a struggle against time. As soon as any contrivance has been made which was unexceptionable as to the mechanism the question has always arisen: Can it not be executed in less time by some other contrivance? Thus every advance has but raised up a new object of rivalry, itself to be superseded by some more rapid means, nor can I hope that I have nearly reached the limit. If I have approached it, it is more than I have a right to expect as the pioneer in this difficult career. Another age must be the judge of that as well as of the other questions relating to the engine. I will only observe that the rapidity of its performance must be judged of first by the velocity which the general structure of the machine enables us to employ with safety. This may in the present engine be greater or less than that I have assigned to it. Secondly, by the number of parts in the time of each cycle which are requisite beyond that necessary for the execution of addition and carriage. Nine units are absolutely necessary for addition and one for carriage and I believe that it may be stated that no engine in which all the digits are added at once can possibly require less than ten units of time for a complete addition or subtraction. Besides this each cycle or turn of the handle must have certain additional units of time allowed it in which the wheels when brought to their right position are locked or rendered immovable and in which also certain motions are given to the axes by which these operations are executed. In the present engine when addition alone is to be executed these extra units amount to five, and when carriage also is required they are ten in number. Thus one of the cycles in which any number can be transferred from one set of figure wheels to another or in which it can be stepped up or down contains fifteen units of time. The other cycle in which addition or subtraction and consequently carriage must be made amounts to twenty units of time.

But it is not merely important that the elements of the operations should be performed in as short a time as possible. The engine ought to be so contrived that in their combination for executing the great operations of multiplication and division the result obtained should be as quickly reached as possible. Several modes have been examined for multiplication and that described has been finally adopted as the shortest.

The reader who only looks at the mechanical difficulties in contriving such an engine will doubtless be surprised when he examines the history of the contrivances, which will be described in a future work, at the lavish rejection of inventions which has taken place in order to achieve rapidity in computation. But the analyst who is aware that the last resource in all our difficulties is the conversion of the most intractable expressions into infinite series will more readily appreciate the importance of even a small abbreviation of the time of an operation, when he is informed that it is one of the objects of this engine to arrive at the numerical value of the coefficients of such series, however complicated the laws by which they are formed, and that proceeding again from these results as etablished data, it is proposed to assign the values of such expressions for any magnitudes of the variables.

In the progress of the invention, it has frequently occurred that mechanical means have either been found altogether insufficient to shorten the time of certain parts of operations, or that it was only possible to effect this object by such great additions of mechanism as were themselves inexpedient.

In this difficulty, I have sometimes had recourse to an arithmetical, or to an algebraical artifice, by which in several instances it has been surmounted without any increase of mechanism. It has sometimes occurred that after the mechanism has been contrived and even drawn, some well-known property of number has enabled me to supersede it. This happened with respect to the means of printing any table *true to the last figure.*

It is well known that though the logarithms of

$$\left.\begin{matrix}2\\3\\4\end{matrix}\right\} \text{ are } \left\{\begin{matrix}.3010299954\\.4771212547\\.6020599913\end{matrix}\right.$$

yet if a table to only seven places of figures is to be printed, the logarithms ought in such a table to be printed:

log of 2 is .3010300
3 .4771213
4 .6020600

In the drawings of the first machine which I contrived for calculating tables by differences there were two arms, one of which after the logarithm had been computed went round to feel if its eighth digit were greater or less than 5. In the former case the arm caused unity to be added to the seventh figure but in the latter case it was inoperative.

After the printing of the seventh figure another arm went in an opposite direction, and subtracted the unity thus added, so that the calculation of the next logarithm might be carried on without impediment from this temporary change. I did not observe until after several contrivances for this purpose had been drawn, that the simple addition of 5 to the eighth figure of the first logarithm computed by the engine would, without any mechanism, ensure the correctness of the seventh figure throughout the whole of the succeeding logarithms of the table.

Another instance of the assistance to be derived from arithmetic occurred in the operation of multiplication. It was found that if the table itself was stepped on its own axes then every alternate turn of the actual multiplication might be saved. The mechanical difficulty which opposed the execution of this plan arose from the necessity of placing those numbers which the digits of the multiplier successively expressed upon the figure wheels of the selecting apparatus. For as this number might be a nine, it was necessary to allow the whole time of one addition, and the movements consequent upon it, or a cycle of fifteen in order to place the number in its proper place; and after this an equal time must be allowed in order to reduce the same wheel to zero, as a preparation for the next multiple. Under these circumstances a remedy without increasing the standard velocity seemed beyond the reach of mechanism.

The following arithmetical artifice removed the obstacle. Let us suppose the multiplier to be the number 584269.[a] If this number be placed on a stepping axis, and stepped down, the lowest figure 9 will be stepped off and lost, but if, instead of this, it be received upon the figure wheel of the selecting apparatus, then the first multiple would be rightly placed. If now a second stepping took place, without obliterating the 9 placed on the selecting apparatus, the number 2 or the second digit of the counting apparatus would be added to the 9, and the result would be a 1, which is not the multiple required.

[a] The number 584269 is sometimes written as 584629 in the text and notations below; since the general argument is unaffected, corrections have not been made.

If this process were continued we should have after each stepping the following corresponding number in the selecting apparatus:

After	1st		2nd		3rd		4th		5th		6th
	5										
	8		5								
	4		8		5						
		on		on		on		on		on	
	6	selecting	4	selecting	8	selecting	5	selecting		selecting	
		wheel		wheel		wheel		wheel		wheel	
	2	9	6	1	4	7	8	1	5	9	4

So that if we do not obliterate the number at each turn, after putting it in the selecting apparatus, the multiplier 584629 will have the same effect as if it had been the multiplier 491719. This led me to enquire what number must be substituted for any given factor, in order that when so stepped, it may produce successively the digits of the given factor.

The question reduced itself to this: *e d c b a* being the digits of a given factor, what must *E D C B A*, the digits of a number to be found be, in order, that by stepping that number on itself, we may have successively the digits of *e d c b a* of the given factor.

The answer to this question gives the equations:

$$A = a$$
$$B = b - a$$
$$C = c - b$$
$$D = d - c$$
$$E = e - d$$

and in the particular case:

$$A = 9,\ B = 7,\ C = 6,\ D = 2,\ E = 4,\ F = 7$$

and if we place the number 742679 on the stepping axis, the successive results will be, after

1st		2nd		3rd		4th		5th		6th	
7											
4		7									
2		4		7							
	on		on		on		on		on		on
6	selecting	2	selecting	4	selecting	7	selecting		selecting		selecting
	wheel		wheel		wheel		wheel		wheel		wheel
7	9	6	6	2	2	4	4	7	8	0	5

Now the same equations inform us how to form the number thus required. For if we place the given multiplier upon a set of figure wheels, and subtract from it, without attending to the carriage of tens, the same number once stepped up, we shall have precisely those figures which are required, thus:

Factor	5 8 4 2 6 9
Factor stepped up	5 8 4 2 6 9
Subtraction without carriage	5 7 4 2 6 7 9

which number, being used instead of the given factor, will produce its successive digits in the figure wheel of the selecting apparatus.

The same artifice has been employed in another part of the engine, for the purpose of overcoming a similar difficulty in the motion of the circular plate carrying the steel types for punching the copperplates.

With respect to the time employed in the operation of multiplication; if p equal the number of digits in P the smallest factor, d equal to the number of decimals in the engine, and if n is equal to the next whole number which is greater than $P/10$; then the time of multiplication without stepping the product is

$$(12 + n) \text{ cycles of } 20 + (p + 4) \text{ cycles of } 15.$$

But if the product is to be stepped to its proper place previously to its being carried into the store, then the time of the operation is

$$(12 + n) \text{ cycles of } 20 + (p + d + 4) \text{ cycles of } 15.$$

These give respectively in turns of the handle

$$\frac{20(12 + n) + 15(p + 4)}{382}$$

for multiplication without stepping, and

$$\frac{20(12 + n) + 15(p + d + 4)}{382}$$

for multiplication with stepping.

6. *Division*

Of the numerous difficulties which the contrivance of an engine to perform arithmetical operations presented, none certainly offered more reasonable grounds of difficulty than that structure of it, by which it should be enabled to perform the operation of division. Contrary to the plan which might perhaps have been expected, I did not commence the attempt with the case of some simple number as 2, 3, or 5 for the divisor, but went at once to the case of large numbers both for divisor and dividend.

The first rough outline of a system by which it could be accomplished offered an instance of the most complicated mechanism I had ever imagined. But before it was submitted to that simplifying process which invariably succeeds the original conception, another principle occurred apparently so simple in its means, that little could be expected beyond it as to principle, and my whole efforts then became directed to the combinations of mechanism, a great part of which already existed in the engine for other purposes.

It has already been stated, that when any number which happens to be greater than another is subtracted from that other number, an event which has been termed a *running up* takes place, and that in consequence of this event, any order which may be desired can be given to any part of the engine. Bearing this fact in mind, the division of one number by another may be conceived to take place in the following manner:

Let the number 00432101234 be divided by 2018.

In turn[a]	Register	
1. Barrel orders divisor to be	00432101234	
subtracted from highest figure	00201800000	
of dividend and to register	00230301234 + 1	
that it has subtracted once		
2. ” ”	00201800000	
	00028501234 + 1	
3. ” ”	00201800000	
	99826701234 + 1	A running up
4. Addition of divisor	00201800000	happens which
Unregister one	00028501234 − 1	directs the addition
5. Step up dividend	0028501234	of the divisor and
6. Barrel orders divisor to be	0020180000	then the stepping
subtracted as before	0008321234 + 1	up of dividend
7. ” ”	0020180000	
	9988141234 + 1	A second running
8. Addition of divisor	0020180000	up and the same
Unregister one	0008321234 − 1	orders as above
9. Step up dividend	008321234	

[a] Several corrections have been made to the diagram. An arithmetic error in turn 26, which should give the result 0050560 instead of 0049560, has not been corrected.

10. Subtract divisor	002018000	
	006303234 + 1	
11. ”	002018000	
	004285234 + 1	
12. ”	002018000	
	002267234 + 1	
13. ”	002018000	
	000249234 + 1	
14. ”	002018000	
	998231234 + 1	Running up
15. Add divisor	002018000	and the same
	000249234 − 1	orders as above
16. Step up dividend	00249234	
17. Subtract divisor	00201800	
	00047434 + 1	
18. ”	00201800	
	99845634 + 1	
19. Add divisor	00201800	
	00047434 − 1	
20. Step up dividend	0047434	
21. Subtract divisor	0020180	
	0027254 + 1	
22. ”	0020180	
	0007074 + 1	
23. ”	0020180	
	9986894 + 1	Running up
24. Add divisor	0020180	and the same
	0007074 − 1	orders as above
25. Step up dividend	0070740	
26. Subtract divisor	0020180	
	0049560 + 1	
27. ”	0020180	
	0029380 + 1	
28. ”	0020180	
	0009200 + 1	
29. ”	0020180	
	9989020 + 1	
30. Add divisor	0020180	
	0009200 − 1	
etc.	009200	

In the above view the first direction given by the barrels is to subtract the divisor from the dividend. This order is repeated until at the end of the third subtraction, it is found that a *running up* has taken place, a fact which of course indicates that the divisor has been subtracted once too often. The number of times the divisor has been subtracted is marked by a register axis. Now it is in consequence of this fact of a running up that the engine *knows* it has subtracted the divisor exactly once too often, and that knowledge is conveyed by the running up lever to the barrels, in the form of an order to move to such verticals, that at the succeeding turns they shall direct the erroneous subtraction and registration to be corrected, and the remainder of the dividend to be stepped up in order to undergo a similar process for the discovery of the next digit in the quotient.

It may perhaps be observed that in the above example another *running up* occurs in the fourth turn at the ninth line in consequence of addition, and it might be expected that an order should consequently follow. But this is provided against by the very general nature of the running up levers, which are capable, under different circumstances, of conveying any order that may be required to any part of the engine. Immediately upon the occurrence of a running up these levers, besides the directions already stated, convey through the barrels an order to themselves to be inefficient at the next running up, thus the runnings up in the 8th, 15th, 19th, 24th, etc., turns are inefficient, whilst in the 3rd, 7th, 14th, 18th, 23rd, and 29th, they are efficient.

The first figure in the quotient being thus obtained, a repetition of the same process will of course produce the succeeding figures.

The complete understanding of this principle is of some importance, for it not merely relates to the process of division, but it is capable of being applied to any tentative arithmetical process. I had applied it to the extraction of roots, but other improvements induced me to dispense with it in that instance. Its application is not limited to existing rules, but I believe I may venture to state, that whenever new methods are contrived for overcoming arithmetical difficulties by tentative processes, it will be found available for those new purposes, and that even in cases where the tentative processes relate to algebraic expressions, it may yet be useful.

I have given a simple case in order to explain the nature of this principle. Division however is not executed by this tedious process, which would require for its performance a number of turns more than equal to the sum of the digits in the quotient. The process really used may be thus described:

The operation card ordering division, having advanced, directs the barrels to the proper verticals.

The divisor P is first received from the ingress wheel upon several parts of the mill.

The dividend R next enters. Several processes are gone through in order to ascertain whether it will be necessary to step the dividend up or down, and how many steppings there ought to be of either kind. This depends upon the number of digits of the quantities operated upon, and is principally executed by the counting apparatus which also computes the number of figures which must occur in the quotient.

At about the eighth turn the formation of the table commences, and also the selecting apparatus is prepared.

If the stepping down is completed, the actual division may commence about the eighteenth turn.

By means of the selecting apparatus the two first figures of the dividend are compared with the two first figures of each of the nine multiples of the divisor, and that multiple which has its two first figures either nearest in magnitude or equal to the two first figures of the dividend is selected and subtracted from the dividend.

The remainder is then stepped up, and the same process repeated until the counting apparatus interferes to finish the process.

In case the two first figures of the dividend are exactly the same as the two first figures of one of the multiples of the divisor, there may arise a doubt whether that multiple or the preceding one should be subtracted; this will depend on the third figure of each quantity. But in this case, the mill always takes for granted that the multiple selected is the true one. If, however, the contrary happens to be the case, then the subtraction of the multiple which is really too large will cause a running up on the dividend, and in consequence of this, orders will be transmitted to the barrels to direct the addition of the first multiple, and the subtraction of unity from the multiple registered in the quotient, by which means the quotient being set right, the first process is then continued.

After the end of the actual division the quotient, and the remainder if required, are given off to the egress wheels.

As the result of a multiplication of a number containing forty digits by another of equal extent is a product containing eighty digits, and as this product can be conveyed into the store on two sets of figure wheels, a provision has been made for bringing such numbers back into the mill when they are to become dividends, so that the result of the division of a number containing eighty by one of forty digits, shall have the forty figures of its quotient true to the last figure.

In deciding on the number of the first figures of the dividend to be compared with those on the table, it was considered that the infrequency of the running up when two only are used, was such that the resulting loss of time would be more than counterbalanced by the mechanism required in case three figures should be compared. If however it should ever be thought expedient to save that portion of time, it is quite possible by adapting a carrying axis to each table axis to render division a direct process, in which no figure can ever be assumed in the quotient which requires a correction afterwards.

The process of division occupies more time than that of multiplication, a circumstance partly arising from the greater extent of the preliminary process, but chiefly from the necessity of occupying one turn of the hand for stepping down, which is executed in multiplication during the time of the addition.

Whenever therefore by proper arrangement of a formula or by other artifices multiplication can be substituted for division time will be gained by the change. Examples of this exchange will be found frequently in the following pages.

The time occupied by the process of division will depend on the numbers concerned, and on the position of the decimal point, and in a small degree on the accidental number of the runnings up.

. . .

7. *Extraction of the square root*

I had at one period in the progress of the engine determined on making it perform the operation of extracting the square root of numbers. For this purpose the usual method was in some degree modified and adapted to the structure of the mill. Advantage was taken of the effects of running up to alter the various directions necessary for the process, and notations were made for the purpose. The time occupied in obtaining each figure was about four turns for every unit it contained, and the process therefore rather long. When successive improvements had considerably diminished the time of the action of the engine, I gave up the process at first intended for the extraction of the square root, and the grounds of the change were:

That with the shortened cycle the process of obtaining the square root by successive approximations would not be longer than the direct extraction in the former state of the engine.

That considerable simplification occurred in the machinery from giving it up.

That the extraction of the cube and all other roots would then be performed according to the same principles.

In the engine now described the extraction of square and other roots will be performed by approximation. If the quantity whose square root is required has a negative sign, the arithmetical root of the number will be found, but a particular notice will be given by which the circumstance that the root so found must be multiplied by $\sqrt{-1}$, will also be indicated. The mode of executing these approximations will be explained in a future page.

IV OF COMPUTING THE NUMERICAL VALUE OF ALGEBRAIC FORMULAE

It will have been observed that the orders by which the engine has been directed to execute the elementary arithmetical operations already described have been of two kinds: operation cards having certain holes by which they govern the barrels, and the variable cards.

These operation cards are always identical with each other, as long as the same operation is to be performed, and they are totally independent on the particular numbers which they may direct to become subject to them. In the cases of multiplication and division, the operation cards will indeed be turned and advanced at the requisition of the barrels in a longer or shorter time according to the nature of the numbers, but the *forms* of these cards will always remain unaltered. In this sense the operation cards partake of that generality which belongs to the algebraic signs they represent.

For the processes of addition and subtraction two operation cards only are necessary. One of these is required for each quantity, and assigns to it the algebraic sign of the process. Thus any number of additions and subtractions are ordered by the same number of operation cards.

Besides these two operation cards it is necessary to have three variable cards. The two first direct the two variables to be operated upon to move from the store into the mill, the third variable card directs some particular store axis to receive the computed sum of the two quantities.

Whatever be the number of quantities connected by the signs plus or minus, the same number of operation cards is required, and the same number increased by one expresses that of the variable cards.

In the processes of multiplication and division, one operation card is required to order the barrels to commence the process. If it should be necessary, under any circumstances, to have two different modes of performing these operations as is at present proposed, for the purpose of saving time in approximations, then another card may be attached, which is to be a blank card, and therefore ineffective in the most frequent case, but which is capable of bringing in the less usual case when it is required.

Besides these operation cards, two variable cards will be required to order in the two factors, and three others to order the three parts of the product into their respective places on the store.

In division three variable cards will be required to order the dividend out of the store into the mill, and one more to direct the quotient to its right place in the store.

The following table exhibits the number of cards necessary for each operation.

	Operation cards	*Variable cards*
To add two quantities	2	3
To subtract one quantity from another	2	3
To multiply two quantities and retain 80 figures of product	1	5
To divide one quantity of 80 figures by another quantity	1	5

In any operation it is only necessary that the operation card ordering its commencement should be advanced. This sets the barrels in action and provides, through the variable cards if necessary, the first quantity to be operated upon; the barrels then call in the other quantities required at the proper times, until the operation is completed, the mill cleared of all numbers, and the computed result placed in its intended situation in the store. The last vertical on the barrels belonging to each operation directs them to move to its zero point.

Such being the arrangement, the calculation of the numerical quantities resulting from two or more successive operations may be readily performed by placing the operation cards strung together in their proper order on their revolving prism, and the variable cards marking the numbers in the store, on which are to be the subject of those operations, and also in the order in which they are required, on their own revolving prism, and then putting on the *final verticals* of one of the barrels a stud which orders the operation cards to advance.

This arrangement being made, after the first advance of the operation cards, the barrels will direct the progress of the operation, calling into the mill by the advance of the variable cards the numbers to be operated upon, and replacing in the store the computed result by the same means. The barrel, by the return to the final verticals of this operation, orders another operation card to advance, which, giving a new order to the barrels, these latter again conduct that operation, demanding at proper times the fit numbers from the store, and conveying back to it their new result, and finally, by returning to their final verticals, they again demand a fresh order from the operation cards.

1. *Verification of the numbers placed in the store*

The numerical quantities intended to be operated upon may be placed by hand on a particular set of figure wheels in the store. Or they may be punched previously by a separate machine on number cards, and thus conveyed into the store.

In either case the possibility of error exists, and although it might reasonably be required that no care should be spared in inserting in the engine correct data for the question whose solution is demanded, yet it is important that the fulfilment of this condition should not depend *solely* on the superintendent of the engine, but that any mistake he might make should be open to detection when the calculated results are returned to those who supplied the formula and constants with which they were to be computed.

Fortunately this condition, which really becomes of great importance when the number of the given constants amount to hundreds or thousands, can be easily fulfilled.

It is merely necessary to order by means of cards that each constant, as it enters the engine, shall pass through the printing apparatus previously to its taking its intended place in the store. By this means the numbers really put into the engine by the attendants will be printed and they may be returned with the computed results printed also by the engine upon the same paper. If the calculation consist of the reduction of a multitude of observations it may be convenient to print, in an adjacent column, each observation entered into the engine and its corrected result.

Thus whatever may be the formula ordered to be computed, the employer of the engine will always possess ample means of detecting any error in the constants inserted in it by the attendant.

2. *Verification of the formulae placed on the cards*

The methods which have been contrived for accomplishing this object are not so simple as that for insuring the accuracy of the constants, nor are they, without the use of considerable machinery, so absolutely certain in their results.

It must, however, be observed, that if care is demanded from the attendants for the insertion of numbers which are changed at every new calculation of a formula, any neglect would be absolutely unpardonable in combining the proper cards in proper order, for the much more important purpose of constructing the formula itself, the arrangement of whose cards is never changed at any after time. This verification might therefore be reasonably left to the diligence of the superintendent.

There are however several subsidiary modes by which he may himself check his own proceedings.

The operation cards, for example, may be made of four different colours corresponding to the four operations of arithmetic:

a white card indicating addition
a yellow card indicating subtraction
a blue card indicating multiplication
a green card indicating division.

When therefore the chain of operation cards is finished, their total number ought to be equal to that of the operations in the formula. The number of white cards ought to agree with the number of additions, the number of yellow cards with that of the subtractions, whilst the number of blue cards marks the number of multiplications, and that of green cards the number of divisions.

Since the number of quantities which enter the mill to be operated upon, as well as that of the results to be returned to the store, is known, the number of the variable cards is known. These might be coloured like the operation cards, and those cards which ordered numbers out of the mill into the store might be coloured balck. If this were the case, before any two sets of formula cards were used they might be compared as to the succession of their colours, when, if correctly placed, it would be found that the same succession of colours would occur in each, but that in the variable cards each colour would be separated from the preceding by one or more black cards.

Another mode of partially verifying a formula consists in assigning such numerical values to the constant quantities as shall render its value easily computed by the pen. If either the operation or the variable cards have been wrongly put together, the formula given to the engine will be different from that which was intended, and the numerical results will in all probability be totally different from that computed by the pen. If trials of three or four simple cases have been made, and are found to agree with the results given by the engine, it is scarcely possible that there can be any error amongst the cards.

When the formula to be computed is very complicated, it may be algebraically arranged for computation in two or more totally distinct ways, and two or more sets of cards may be made. If the same constants are now employed with each set, and if under these circumstances the results agree, we may then be quite secure of the accuracy of them all.

It should, however, never be forgotten, that whatever attention may have been bestowed on securing the accuracy of the cards, that object, when once attained can never again become doubtful.

If, notwithstanding these methods of verification, it should still be considered desirable to have greater security, it is certainly possible to attain it by making the engine itself print the formula inserted in it, as well as the constants to be employed in that formula.

At two different stages in the progress of the invention, I had contrived mechanism by which this might be accomplished, but as it related to a much more advanced stage, I did not make any drawings of it. Since the introduction of formula cards, these contrivances have become susceptible of great improvement. For it is now no longer necessary to add mechanism for that purpose to an engine already sufficiently complex; but a separate machine should be constructed, by placing the formula cards in which the formula might itself be printed from them.

Such a machine does not appear to me to present much difficulty, as far as regards selecting the signs and variables. The chief inconvenience would arise from the uncertain extent of the compound quantities operated upon. This might in a great measure be obviated by only requiring it to print the series of successive substitutions which are to be made in the engine. The reader will understand the force of these remarks better when he has perused the mathematical part of this volume.

But even this assistance might fail in some of the more complicated cases, for a certain degree of doubt always remains whether we have ourselves worked out rightly the developed results which we wish to convert into number. This doubt may be almost altogether removed when we avail ourselves of the *combinatorial cards*, because in such instances, although the ultimate results are extremely complicated, yet the instructions communicated to the engine by cards are very few, and exceedingly simple, and may therefore be repeatedly verified in a very short time. It appears however to me that the formula printing-machine might by some improvements itself ultimately work out many of such algebraic developments.

3. *Extent of the numerical power of the engine*

In deciding on the extent of the numbers with which the engine should compute, the first consideration was to look at the number of figures which in the present state of mathematical enquiry are required in the most extensive calculations. It then became desirable to look forward to the probable increase which improved observations might require. Finally, the mechanical structure of the engine, or its necessary arrangement, might put limits to this extent, or render the time employed in given calculations longer, at certain definite intervals.

But even if it were thus possible to satisfy all practical wants there would still remain a desideratum to render the mechanism philosophically perfect as to its power of converting algebraic expressions into numbers, without any limit as to their magnitude or extent.

The result of my reflections has been that numbers containing more than thirty places of figures will not be required for a long time to come. I have, however, made the drawings of the engine for forty places of figures. All additions and subtractions may be made with such numbers and the products, amounting to eighty places, may be preserved in the store and brought back into the mill to be divided by numbers of forty places of figures, thus retaining forty places in the quotient.

I have however considered thirty places of figures as the standard, and upon that basis have made all the estimates of the time required for the computation of formulae. In order to fulfil the condition of the conversion of algebraic formulae into number without reference to the number of digits in the quantities employed, I have availed myself of certain arithmetical principles, by the aid of which, without adding to the mechanism, calculations, if requiring any extent of digits, may be performed; these principles will be explained in a future page. At present it may be sufficient to observe that if it is required to compute formulae having in each quantity not more than eighty figures, it will be sufficient to employ about four times the time requisite for the calculation of the same formulae, having only half that number of digits in each quantity. If three times the number of digits (or 120 digits) be required in each number, then about nine times the time will be requisite and in general, if the number of figures required to be used in the given quantities be expressed as n times the number forty, then the time required by the engine will be: $40 \times n^2$.

It was important, at least for the theoretical perfection of the engine, that no additional mechanism should be required for any extent of numerical calculation, beyond that which was considered to be safely practicable in the present state of the mechanical arts.

This point has been accomplished with respect to numbers, by increasing the time nearly in proportion to the square of the number of times the sum of the digits is a multiple of forty.

As to the extent in point of successions of the algebraical operations, no practical limit seems to exist. The number of cards which indicate the operations, and those that govern the quantities, may be increased to any extent without loading the engine, and in the more complicated developments the combinatorial cards may be made use of, and expansions, which by the usual means would require above twenty thousand cards, may be performed with little more than two hundred.

4. *Time required in computing formulae*

The most constant difficulty in contriving the engine has arisen from the desire to reduce the time in which the calculations were executed to the shortest which is possible.

It is not to be presumed that such an attempt has succeeded. How near the approach has been made must remain for aftertimes to determine. It is, however, desirable to state the mode in which the time has been estimated and to point out sources of mistake which might perhaps be overlooked by those who have not deeply studied the subject.

The first step is to assume as a definition, the unit of space and the unit of time. I have assumed for the first of these, the space (measured on a circle), between the figures on the figure wheel. This is ____. The unit of time is assumed to be that portion of a second required for the passage of one figure to another on the same wheels.

The time occupied by one turn of the handle of the engine, which is called a cycle, consists of the time necessary for the directive action, the time requisite for executing those directions, the time necessary for addition, the time necessary for locking the wheels moved, the time necessary for carriage, and the time required for locking the wheels after carriage.

It sometimes occurs that carriage is not required, in this case the cycle consists of fifteen units of time. When carriage also is to be used, twenty units of time are necessary.

The first mode of reducing the time which presents itself is by diminishing the unit of space, or the distance between the figures. The first inconvenience which presents itself is that the machinery becomes more delicate and requires better workmanship. The general scale on which it is constructed becomes that of clockwork, and finally is reduced to watchwork. This evil will of course be diminished as the mechanical arts advance in perfection.

Another consequence of the diminution of the unit of space is, that as the *time lost* at each gearing is not proportionally diminished, it becomes necessary to make more frequent corrections by means of additional locking apparatus. The time required for these lockings increases the length of the cycle . . . the unit of space was assumed to be ·62832 and a correction was applied to destroy the loss of time at every third wheel. When the two counteracting principles of a short unit and many lockings were distinctly perceived the present unit was adopted as that most fitted for the present state of mechanical art and a locking was placed at every second wheel.

The time necessary for the action of the directive part consists in that employed in advancing the barrels, and in turning and advancing the several classes of cards. The advance of the barrels and that of the cards can be made contemporaneous, and the extent of their motion will depend on the mode by which the several parts are thrown into gear, and the time will depend upon the space, and the mass of matter to be moved.

The time necessary for the circular motion of the barrels from one vertical to another, and also that required for the turning over several cards to arrive at the one wanted, when it is at some distance from that last used, must also be considered. This operation may be executed during the addition part of the cycle, provided the distance between the vertical in the barrels and the cards on the prism is not too large. When the unit of space ·62832 was employed, fourteen verticals could be passed over in a cycle and ____ cards. But when that unit was reduced to ____ it was found that only 5 and 7 verticals would be passed and ____ cards in the same portion of the cycle. The consequence of this diminished time will be that under certain circumstances it may become necessary to employ an extra turn of the handle in order to bring the barrels or the cards to their proper positions. This then forms the first counteracting principle to be considered when it is proposed to shorten the unit of space.

The corrections for loss of time arising from the original structure and subsequent wear of the teeth of the wheels are made by means of the locking apparatus. Levers terminating in wedges are forced between the inclined planes of the locking teeth, and thus bring the wheels to which they are attached accurately to their places. The depth of the locking teeth depends on the extent of the correction, and the time occupied in moving them will depend partly on that depth, and partly on the clearance required when they are unlocked. If therefore the same extent of motion is carried through a short and through a long train of wheels, the time occupied in bringing the last wheel to its true place will in the first case arise from few lockings moving through certain spaces, whilst in the long train it will result from more numerous lockings, each of which moves through a smaller space.

If the space required for clearance could with safety be diminished in the same proportion as the depth of the lockings, then in point of time merely, it would be indifferent whether the lockings were few or many, and their number would then be determined by the simplicity of mechanism, and by other principles. The unlocking of the several wheels may take place in a much shorter time than that required for locking them.

The time requisite for executing the orders of the directive part consists in that during which the original source of movement is connected with the several parts which are ordered to be moved. The time of addition comes within this class, and so also does that in which certain gearings are made and broken by the lifting or descent of certain axes. As these gearings can usually be made at the same moment, / their number does not augment the time they employ. But if, in order to save mechanism, those axes move to several points of rest and consequently pass over considerable space, then more time must necessarily be allowed. When the number of figure wheels in each cage was reduced from four to two, the time employed in lifting these axes was also reduced.

The time consumed in carriage varies more than that of any other portion of the cycle, according to the different mechanism employed for that purpose. The mere time of adding the units caused by carriage may occupy one unit of time or it may require forty. The time for the previous adjustment will also vary but in a less degree. It is therefore on this point that by far the most important saving can be effected, a saving which would justify even very expensive machinery.

In the earlier stages of the invention of the engine, the great extent of the mechanism for carriage, which it was necessary to attach to each set of figure wheels, became a serious obstacle to its advancement. It was the pressure of this difficulty which led me to reflect on the great assistance derived in analysis, from separating the symbols of operation from those of quantity, and by a distant analogy to attempt the introduction of a carrying apparatus, which might be attached, as occasion required, to any set of figure wheels.

When I reflected on the fact that the quantity of mechanism necessary for performing the operation of mere addition without carriage was not above the tenth part of that requisite for carriage itself, the importance of such a separation became evident, and this led me to separate the machine into two parts, one (the mill) in which operations were performed the other (the store) in which the quantities used and produced were inserted.

In its present state the time by which the cycle is increased when carriage is required is five units, only one of which, however, is employed in the actual carriage. The rest of the five is filled up by the necessary motions for preparing carriage, and for the additional lockings consequent upon it.

The introduction of the principle of hoarding carriages until they amount to nine, has in the case of multiplication enabled me to reduce that operation still further by substituting *nine short* and *one long* cycle, for every *ten long* cycles previously employed.

The time necessary for locking the wheels after carriage will be nearly the same as that requisite for locking them before. According to the nature of the mechanism a gearing or two more or less will make the only difference.

Previously to deciding on the absolute time of each cycle, another question must be entered upon. Since the most important circumstance which affects the duration of a cycle by a time amounting, even in the most favourable circumstances yet contrived, to one third of that in which addition can be performed, is the time which is required for carriage, it may reasonably be enquired whether, by adopting a different radix for the arithmetical notation, we may not in some measure avoid the loss of time. I accordingly at one period considered the advantage of adopting the arithmetical notation of which the base is 100 instead of 10.

The immediate consequence of such a plan was that carriage occupied on the average about one ninth of the usual time. Some drawings were made for this view of the subject but after much consideration the plan was given up as undesirable even in an engine of differences.

Having thus examined the elements of the operations, and pointed out the chief ingredients of the time they occupy, the next consideration is their combination into operations. As far as I have been able to examine the processes of addition and subtraction, it does not appear that much is to be gained, since the lowest limit must be one turn for each process and the limit is at present $n + 4$ turns for n operations. The small saving which might possibly be contrived is not of the same importance as any economy in the time of performing the larger operations.

With respect to multiplication two methods may be followed, either the process may be commenced by taking that multiple out of the table which corresponds to the lowest or units figure of the multiplier, and then proceeding to the multiple corresponding to the tens, and so on; or we may begin by selecting the multiple corresponding to the highest figure, then that to the second figure and so on.

The first of these modes of proceeding requires only one carriage, because at each stepping down of the product one figure is carried off to another axis, and no subsequent addition to the product can ever alter the figures so transferred. If, however, we commence at the highest figure of the multiplier two carriages become necessary, because any multiple, down even to that which the units figure directs, may by possibility change all the others up to the highest, and since the product of forty figures by forty produces a number of eighty places of figures, this must be placed upon two sets of figure wheels and consequently require two carriages.

The operation of division does not appear to admit of the same two methods as those which are applicable to subtraction. It appears to be necessary to perform subtraction, and then step up the remainder, and any attempts I made to unite the two processes so as to perform them in one cycle have proved ineffectual.

. . .

The time occupied by the engine in computing the same formulae will be different according to the mode in which it is arranged. This will be apparent from a single case, let

$$x = \frac{a^2c + b^2a + c^2b}{abc}.$$

This will, in its present form, require eight multiplications, two additions, and one division.

If, however, it be put into the form

$$x = \frac{a}{b} + \frac{b}{c} + \frac{c}{a}.$$

it will only require three divisions, and two additions. In the latter case the time of computing it will be but little more than half that required in the former.

The difference is still greater in other cases. If it is required to find the arithmetical value of

$$a^6 + 5a^5b + 10a^4b^2 + 20a^3b^3 + 10a^2b^4 + 5ab^5 + b^6$$

each term will require five multiplications besides that which arises from its numerical coefficient, so that even ommitting the time so occupied, the computation will require thirty-five multiplications and six additions whilst if it be put under the form

$$(a + b)^6$$

the same calculation may be made by five multiplications and one addition.

The consequences resulting from this circumstance are important, and deserve the attention of those who are engaged in extending the domain of analysis, as well as of those who look forward to the effects which are likely to be produced by the complete control which mechanism now gives us over number.

Whenever engines of this kind exist in the capitals and universities of the world, it is obvious that all those enquirers who wish to put their theories to the test of number, will apply their efforts so to shape the anlytical results at which they have arrived, that they shall be susceptible of calculation by machinery in the shortest possible time, and the whole course of their analysis will be directed towards this object. Those who neglect the indication will find few who will avail themselves of formulae whose computation requires the expense and the error attendant on human aid.

[352]

V

NOTIONS SUR LA MACHINE ANALYTIQUE DE M. CHARLES BABBAGE[a]

PAR L. F. MENABREA
(Capitaine du génie militaire)

Bibliothèque universelle de Genève, Vol. 41
(1842), pp. 352–76

Les travaux qui appartiennent à plusieurs branches des sciences mathématiques, quoique paraissant, au premier abord, être uniquement du ressort de l'esprit, peuvent néanmoins se diviser en deux parties distinctes: l'une qu'on peut appeler mécanique, parce qu'elle est sujette à des lois précises et invariables, susceptibles d'être traduites physiquement, tandis que l'autre qui exige l'intervention du raisonnement, est plus spécialement du domaine de la pensée. Dès lors on pourra se proposer de faire exécuter par le moyen de machines la partie mécanique du travail, et réserver à la seule intelligence celle qui dépend de la faculté de raisonner. Ainsi la rigueur à laquelle sont soumises les règles du calcul numérique a dû, depuis longtemps, faire songer à employer des instruments matériels, soit pour exécuter entièrement ces calculs, soit pour les abréger. De là sont nées plusieurs inventions dirigées vers ce but, mais qui ne l'atteignent, en général, qu'imparfaitement. Ainsi la machine de Pascal, tant vantée, n'est maintenant qu'un simple objet de curiosité qui, tout en prouvant une /

[a] This paper is the original French version of L. F. Menabrea's account of the Analytical Engine which was translated by Ada Lovelace in her 'Sketch of the Analytical Engine' (1843), and which begins on p. 89 in this volume of the *Works of Babbage*.

grande force d'intelligence dans son inventeur, présente en elle-même peu d'utilité. Elle n'exécutait que les quatre premières opérations de l'arithmétique, et encore se bornait-elle, en réalité, aux deux premières, car la multiplication et la division étaient le résultat d'une série d'additions et de soustractions. Le grand inconvénient de la plupart de ces machines est d'exiger l'intervention continuelle de la main de l'homme pour en régler les mouvements, et c'est en cela qu'est la source des erreurs; de sorte que, si dans l'exécution des grands calculs numériques leur usage ne s'est pas propagé, c'est qu'elles ne résolvaient point le double problème que présente la question, celui d'obtenir l'exactitude des résultats unie à l'économie du temps.

Frappé de ces réflexions, Mr Charles Babbage a consacré plusieurs années à réaliser une pensée gigantesque. Il ne s'est proposé rien moins que de construire une machine capable d'exécuter, non-seulement les calculs arithmétiques; mais encore les calculs analytiques, dont les lois seraient connues. L'imagination est d'abord effrayée d'une telle entreprise, mais à mesure que l'on réfléchit avec plus de calme, le succès en paraît moins impossible, et l'on sent qu'il peut dépendre de la découverte de quelque principe, assex général pour que, si on l'applique à la machine, celle-ci soit capable de traduire mécaniquement les opérations qui lui seraient indiquées par l'écriture algébrique. L'illustre inventeur, lors d'un voyage qu'il fit à Turin, ayant bien voulu me communiquer quelques unes de ses considérations à cet égard, je viens, avec son consentement, rendre compte des impressions qu'elles ont produites dans mon esprit. On ne doit point s'attendre à trouver ici une description de la machine de Mr Babbage: elle exigerait de longues études pour être comprise, mais je tâcherai d'en faire saisir le but et d'exposer les principes sur lesquels est fondée son exécution.

D'abord je dois prévenir qu'elle est entièrement différente de celle dont on trouve un aperçu dans le traité de l'Economic / des machines du même auteur. Mais, comme cette dernière a fait naître la pensée de la machine en question, je crois qu'il est utile, avant tout, de rappeier en peu de mots quelles ont été les premières tentatives de Mr Babbage, et quelles sont les circonstances qui y ont donné lieu.

On sait que le gouverenement français, voulant faciliter l'extension du système décimal, avait ordonné de construire des tables logarithmiques et trigonométriques d'une étendue immense. Mr de Prony, qui avait été chargé de la direction de cet ouvrage, le répartit en trois sections, à chacune desquelles furent appliqués des hommes spéciaux. Dans la première section l'on combinait les formules de manière à les rendre propres aux calculs arithmétiques; dans la deuxième, ces mêmes formules étaient calculées pour des valeurs de la variable, prises de distance en distance; enfin dans la troisième section, composée de près de 80 individus, qui pour la plupart ne connaissaient que les deux premières règles de l'arithmétique, on interpolait, au moyen de simples additions ou soustractions, des valeurs intermédiaires à celles qui étaient calculées par la 2e sect.

Un travail analogue à celui que nous venons de citer ayant dû se répéter en Angleterre, Mr Babbage pensa que les opérations faites par la troisième section pouvaient être exécutées par une machine, et c'est ce qu'il a réalisé au moyen d'un mécanisme dont une partie a été confectionnée, et que l'on peut appeler *machine aux différences* à cause du principe sur lequel elle est fondée. Pour en donner une idée, il suffira de considérer la série des nombres carrés entiers: 1, 4, 9, 16, 25, 36, 49, 64, etc. En faisant la différence de chacun de ces nombres avec le suivant, on aura une nouvelle série que nous appellerons série des différences premières, composée des nombres: 3, 5, 7, 9, 11, 13 15, etc. Si l'on soustrait de chacun de ces nombres son précédent, l'on obtiendra les différences deuxièmes, qui sont toutes constantes et égales à 2. La suite de ces opérations et des résultats auxquels elles conduisent peut être représentée par le tableau suivant: /

	A	*B*	*C*	
	Colonue des nombres carrés	*Différences premières*	*Différences deuxièmes*	
	1			
		3		
	4		2	*b*
		5		
a	9		2	*d*
		7		
c	16		2	
		9		
	25		2	
		11		
	36			

D'après la manière dont les deux dernières colonnes B et C ont été formées, il est facile de voir que, voulant, par exemple, passer du nombre 5 au suivant 7, il faudra ajouter au premier la différence constante 2; de même, si du carré 9 on veut passer au suivant 16, on ajoutera au premier la différence 7, ou en d'autres termes, la différence précédente 5, plus la différence constante 2; ou bien encore, ce qui revient au même, pour avoir le nombre 16, il suffira de sommer ensemble les trois nombres 2, 5 et 9, placés dans la direction oblique *a b*. D'une manière analogue, on obtiendra le nombre 25 en sommant ensemble les trois nombres placés dans la direction oblique *d c*: en commençant par la somme 2 + 7, on a la différence première 9, consécutive à 7; en y ajoutant 16, on aura le carré 25. On voit donc, qu'étant donnés les trois nombres 2, 5 et 9, toute la série des nombres carrés suivants, ainsi que celle de leurs différences premières, pourra s'obtenir par de simples additions. Maintenant, pour concevoir comment ces opérations peuvent être reproduites par une machine, supposons que celle-ci ait trois cadrans que nous désignerons par *A*, *B*, *C*, et sur chacun desquels soient tracées mille divisions, par exemple, qui seront parcourues par une aiguille. Les deux cadrans *C* et *B* auront, en outre, chacun une sonnerie qui battra un nombre de coups égal à celui des divisions indiquées par l'aiguille. Pour chaque coup du compteur du cadran *C* l'aiguille *B* devra avancer d'une division; il en sera de même de l'aiguille *A*, qui avancera d'une division pour chaque coup du cadran *B*. Telle est la disposition / générale du mécanisme. Cela posé, au commencement de la série d'opérations que nous voulons exécuter, mettons l'aiguille *C* surm la division 2, l'aiguille *B* sur la division 5, et l'aiguille *A* sur la division 9. Faisons sonner le compteur du cadran *C*: il battra deux coups, et en même temps l'aiguille *B* marchera de deux divisions. Alors celle-ci marquera le nombre 7, qui suit le nombre 5 dans la colonne des différences premières. Faisons sonner à son tour le compteur du cadran *B*: il battra 7 coups, pendant lesquels l'aiguille *A* avancera de 7 divisions qui, ajoutées au nombre 9 qu'elle marquait déjà, donneront le nombre 16, soit le carré consécutif à 9. En recommençant l'opération à partir de l'aiguille *C*, qu'on laissera toujours sur la division 2, l'on voit qu'en la continuant d'une manière indéfinie, on pourra reproduire la série des carrés des nombres entiers au moyen d'un simple mécanisme.

Le théorème sur lequel est fondée la construction de la machine que nous venons de décrire, est un cas particulier d'un autre théorème plus général, et qui consiste en ce que si, dans un polynome quelconque dont la plus haute puissance de la variable soit *m*, l'on fait croitre cette même variable par degrés égaux, que l'on calcule les valeurs correspondantes du polynome, puis, que l'on prenne successivement leurs différences première, deuxième, troisième, etc., ainsi qu'il a été fait pour la série des carrés, les différences $m^{ième}$ seront toutes égales entre elles. De sorte que, pour reproduire la série des valeurs du polynome au moyen d'une machine analogue à la précédente, il suffira qu'elle ait $(m + 1)$ cadrans ayant les uns avec les autres les relations indiquées. Comme les différences peuvent être positives ou négatives, la machine contiendra un mécanisme qui fera avancer ou reculer l'aiguille, selon que le nombre à ajouter algébriquement aura le signe *plus* ou le signe *moins*.

Lorsque d'un polynome on passe à une série d'une infinité de termes ordonnés suivant les puissances ascendantes de la variable, pour appliquer la machine au calcul de la fonction représentée par cette série, il semblerait au premier abord que / le mécanisme devrait contenir une infinité de cadrans, ce qui rendrait la chose impossible. Mais la difficulté disparaîtra pour bien des cas, si on observe que, pour un grand nombre de fonctions, l'on parvient à rendre convergentes les séries qui les représentent, de sorte que, selon le degré d'approximation que l'on désire, on pourra se bourner à ne calculer qu'un petit nombre des termes de la série, les autres pouvant être négligés. De cette manière la question est ramenée au cas primitif d'un polynome fini. C'est ainsi que l'on peut calculer la suite des logarithmes des nombres. Mais comme, à mesure que la variable augmente de valeur, les termes que l'on avait d'abord négligés prennent un accroissement qui finirait par influer sur le degré d'approximation que l'on veut conserver, il faut, à certains intervalles calculer la valeur de la fonction par des procédés divers, et partir de ces résultats pour en déduire, par le moyen de la machine, les autres valeurs intermédiaires. Comme on le voit, cette machine remplace ici la troisième section des travailleurs dont il a été parlé au sujet des tables calculées par ordre du gouvernement français, et elle remplit ainsi le but proposé.

Telle est la première machine imaginée par Mr Babbage. On voit que son emploi est limité aux cas où les nombres demandés sont susceptibles d'être obtenus par de simples additions ou soustractions, qu'elle n'est pour ainsi dire que l'expression d'un théorème particulier d'analyse, et qu'enfin elle ne s'étend point à la solution d'une infinité d'autres questions qui sont du ressort de l'analyse mathématique. C'est en songeant au vaste champ qui lui restait encore à parcourir, que Mr Babbage, renonçant à ses premiers essais, conçut le plan d'un autre système de mécanisme dont l'usage devait avoir la généralité de l'écriture algébrique même, et que, pour cette raison, il nomme *machine analytique.*

Maintenant que j'ai exposé l'état de la question, il est temps de faire connaître le principe sur lequel est fondée la construction de cette machine. Lorsqu'on emploie l'analyse pour résoudre / quelque problème, il y a, en général, deux espèces d'opérations à exécuter: d'abord le calcul numérique des différents coefficients, et ensuite leur distribution par rapport aux quantités qui doivent en être affectées. Si l'on a, par exemple, à former le produit de deux binomes $(a + bx)(m + nx)$ le résultat sera représenté par $am + (an + bm)x + bnx^2$, expression dans laquelle on devra d'abord calculer *am*, *an*, *bm*, *nb*, sommer ensuite $an + bm$, et enfin distribuer les coefficients ainsi obtenus par rapport aux puissances de la variable. Si l'on veut reproduire ces opérations au moyen d'une machine, elle devra donc posséder deux facultés distinctes: 1° cella d'exécuter les calculs numériques, 2° celle de distribuer convenablement les valeurs ainsi obtenues. Mais si la main de l'homme était obligée d'intervenir pour diriger chacune de ces opérations partielles, il n'y aurait rien de gagné sous le rapport de l'exactitude et de l'économie de temps; il faudra donc encore que la machine ait la propriété d'exécuter ellemême toutes les opérations successives nécessaires pour la solution du problème qu'on lui propose, une fois qu'on y aura introduit les *donnèes numèriques primitives* de ce même problème. Ainsi, puisque, une fois que la nature du calcul à exécuter ou celle du problème à résoudre sont indiquées, la machine doit, en vertu de son propre pouvoir, passer d'elle-même par toutes les opérations intermédiaires qui conduisent au résultat proposé, elle exclura toutes les méthodes de tâtonnement et d'essai, et n'admettra que les procédés directs de calcul. Il doit en être ainsi, car la machine n'est point un être qui pense, mais un simple automate qui agit suivant les lois qu'on lui a tracées. Cela posé, une des premières recherches de l'auteur a dû être celle d'un moyen de faire la division d'un nombre par un autre, sans employer la méthode de tâtonnement indiquée par les règles ordinaires de l'arithmétique. Cette combinaison n'a pas été la moins difficile à obtenir; c'est d'elle que dépendait le succès de toutes les autres. Dans l'impossibilité où je suis / de décrire ici le procédé par lequel on arrive à ce but, on devra se borner à admettre que les quatre premières opérations de l'arithmétique, c'est-à-dire l'addition, la soustraction, la multiplication et la division, sont susceptibles d'être exécutées d'une manière directe par l'intermédiaire de la machine. Cela étant, la machine sera par là même capable de faire toute espèce de calcul numérique, car en définitive ces calculs se réduisent tous aux quatre opérations que nous venons d'énoncer. Pour concevoir maintenant comment la machine peut fonctionner suivant les lois établies, donnons d'abord une idée de la manière dont les nombres y sont matériellement représentés.

Figurons-nous une pile ou colonne verticale composée d'un nombre indéfini de disques circulaires, tous traversés dans leur centre, par un axe commun autour duquel chacun d'eux peut prendre un mouvement de rotation indépendant. Si, sur le contour de chacun de ces disques, on écrit les dix chiffres qui composent notre alphabet numérique, en disposant, suivant une même verticale, une série de ces chiffres, on pourra exprimer de cette manière un nombre quelconque. Car il suffira de supposer que le premier disque représente les unités, le deuxième les dizaines, le troisième les centaines, ainsi de suite. Lorsque deux nombres auront été écrits de cette manière sur deux colonnes distinctes, on pourra se proposer de les combiner aithmétiquement entre eux, et d'obtenir le résultat inscrit sur une troisième colonne. En général, si l'on a une série de colonnes composées de disques et que nous désignerons par V_0, V_1, V_2, V_3, V_4, etc., on peut demander, par exemple, de diviser le nombre écrit sur la colonne V_1, par celui de la colonne V_4, et d'obtenir le résultat sur la colonne V_7. Pour faire cette opération, il faudra donner à la machine deux dispositions distinctes: la première par laquelle elle est prête à exécuter *une division*, et la seconde par laquelle on lai indique les colonnes sur lesquelles elle doit opérer, et celle où le résultat doit être écrit Lorsqu'à cette division devra, par exemple, succéder une / addition de deux nombres pris sur d'autres colonnes, les deux dispositions de la machine devront changer à la fois. S'il s'agit, au contraire, d'exécuter une suite d'opérations de la même nature, alors la première disposition demeurera la même, la seconde seule changera. Ainsi, parmi les arrangements que peuvent prendre les diverses pièces de la machine, on en distinguera deux principaux: 1° *la disposition relative aux opérations*, 2° *la disposition relative aux variables;* par cette dernière on entend celle qui indique les colonnes sur lesquelles on doit opérer. Quant aux opérations mêmes, elles s'exécutent dans un appareil spécial désigné sous le nom de *moulin;* il est lui-même composé d'un certain hombre de colonnes semblables à celles des variables. Lorsque deux nombres doivent être combinés ensemble, la machine commence par les effacer sur les colonnes où ils étaient écrits, c'est-à-dire à mettre *zèro* sur chaque disque dans les deux lignes verticales où les nombres étaient écrits; puis elle les transporte dans le moulin. Là, l'appareil étant disposé pour l'opération demandée, celle-ci s'exécute, et, une fois qu'elle est achevée, le résultat est lui-même transporté sur la colonne des variables qui aura été indiquée. Ainsi le moulin est la partie de la machine qui exécute; les colonnes des variables constituent celle où s'écrivent et s'ordonnent les résultats. D'après ce qui vient d'êcrivent et s'ordonnent les résultats. D'après ce qui vient d'être dit, l'on voit que tous les résultats fractionnaires et irrationnels seront représentés par des fractions décimales. En supposant chaque colonne composée de quarante disques, cette extension serait suffisante pour tous les degrés d'approximation dont on a généralement besoin.

Maintenant l'on demandera comment la machine peut d'ellemême et sans avoir recours à la main de l'homme, prendre les dispositions successives convenables pour opérer. La solution de ce problème a été empruntée à l'appareil de Jacquard, en usage pour la confection des étoffes brochées; voici de quelle manière:

Dans les tissus on distingue généralement deux espèces de / fils: d'abord, les chaînes ou fils longitudinaux, ensuite la trame ou fil transversal qui est guidé par l'instrument que l'on nomme navette et qui croise avec les chaînes. Lorsque l'on veut faire une étoffe brochée, il faut empêcher tour à tour certains fils de croiser avec la trame, suivant un ordre déterminé par la nature du dessin qu'il s'agit de reproduire. Jadis cette opération était longue et difficile et exigeait que l'ouvrier, attentif au desin qu'il devait copier, ordonnât lui-même les mouvements que ces fils devaient prendre. De là provenait le prix élevé de ce genre d'étoffes, surtout lorsqu'il y entrait des fils de différentes couleurs. Pour simplifier cette manufacturation, Jacquard imagina de faire communiquer chacun des groupes de fils qui devaient agir ensemble avec un levier distinct pour chaque groupe. Tous ces leviers se terminent par des tiges réunies en un faisceau qui a généralement la forme d'un parallélipipède à base rectangulaire; ces tiges sont cylindriques et séparées entre elles par de petits intervalles. L'opération du soulèvement des fils se réduira ainsi à mouvoir convenablement ces divers bras de levier. Pour cela, on prend une feuille de carton de forme rectangulaire, un peu plus grande que la section du faisceau. Si l'on applique cette feuille contre la base du faisceau de leviers, en lui imprimant un mouvement de translation, ceux-ci se mouvront tous en même temps, ainsi que les fils qui leur correspondent. Mais si le carton, au lieu d'être *plein*, était percé dans les endroits où les extrémités des leviers viennent le rencontrer, alors, dans son mouvement le carton étant traversé de part en part par les leviers, ceux-ci resteraient en place.

De cette manière on voit qu'il est aisé de déterminer la position de ces trous dans le carton, de manière que, à un instant donné, il y ait un certain nombre de leviers et par conséquent de fils soulevés, tandis que les autres restent en place. En supposant que cette opération soit successivement répétée suivant une loi indiquée par le dessin qu'on veut exécuter, l'on conçoit que ce dessin puisse être reproduit sur l'étoffe. Pour cela il n'y a qu'à composer une série de cartons suivant la loi requise, et à / les disposer convenablement à la suite les uns des autres; puis en les faisant passer sur an arbre à section polygonale, qui à chaque coup de la navette présenterait une nouvelle face, laquelle serait transportée parallèlement à elle-même contre le fasisceau de leviers, l'opération du soulèvement des fils se ferait d'une manière régulière. On voit qu'ainsi l'on pourra confectionner les étoffes brochées, avec une rapidité et une précision qu'il était difficile d'obtenir auparavant.

Ce sont des dispositions analogues à celle que nous venons de décrire, qui ont été introduites dans la machine analytique. Elle contient deux espèces principales de cartons: 1° les cartons des opérations, par lesquels la machine est disposée de manière à exécuter une série déterminée d'opérations, telles que additions, soustractions, multiplications et divisions; 2° les cartons des variables, qui indiquent à la machine des colonnes sur lesquelles les résultats doivent être écrits. Les cartons étant mis en mouvement, disposent successivement les diverses pièces de la machine selon la nature des opérations à faire, et la machine les exécute en même temps au moyen des mécanismes dont elle est composée.

Pour mieux concevoir la chose, prenons pour exemple la résolution de deux équations du premier degré à deux inconnues. Soient donc les deux équations suivantes où x et y sont les inconnues:

$$\begin{cases} mx + ny = d \\ m'x + n'y = d' \end{cases}$$

on en dēduit

$$x = \frac{dn' - d'n}{n'm - nm'}$$

et pour y une valeur analogue. Représentons toujours par V_0, V_1, V_2, etc., les différentes colonnes qui contiennent les nombres, et supposons qu'on ait choisi les huit premières colonnes pour y écrire les nombres qui sont représentées par m, n, d, m', n', d', n et n', ce qui suppose $V_0 = m$,

$V_1 = n$, $V_2 = d$, $V_3 = m'$, $V_4 = n'$, $V_5 = d'$, $V_6 = n$, $V_7 = n'$. La série des opérations commandées par les cartons, et les résultats obtenus, seront représentés par le tableau suivant: /

Nombre des opérations	*Cartons des opérations*	*Cartons des variables*		*Marche des opérations*
	Signes indiquant la nature des opérations	*Colonne soumise aux opérations*	*Colonne recevant le résultat des opérations*	
1	×	$V_2 \times V_4 =$	V_8	$= dn'$
2	×	$V_5 \times V_1 =$	V_9	$= d'n$
3	×	$V_4 \times V_0 =$	V_{10}	$= n'm$
4	×	$V_1 \times V_3 =$	V_{11}	$= nm'$
5	−	$V_8 - V_9 =$	V_{12}	$= dn' - d'n$
6	−	$V_{10} - V_{11} =$	V_{13}	$= n'm - n'm'$
7	÷	$\dfrac{V_{12}}{V_{13}} =$	V_{14}	$= x = \dfrac{dn' - d'n}{mn' - m'n}$

Comme les cartons ne font qu'indiquer comment et sur quelles colonnes la machine doit agir, il est clair qu'il faudra encore, dans chaque cas particulier, introduire les données numériques du calcul. Ainsi, dans l'exemple que nous avons choisi, on devra préalablement écrire les valeurs numériques de m, n, d, m', n', d' dans l'ordre et sur les colonnes indiquées, après quoi l'on fera agir la machine qui donnera la valeur de l'inconnue x pour ce cas particulier. Pour avoir la valeur de y, il faudra faire une autre série d'opérations analogues aux-précédentes. Mais on voit qu'elles se réduiront à quatre seulement, car le dénominateur de l'expression de y, sauf le signe, est le même que celui de x, et égal à $m'n - mn'$. Dans le tableau précédent l'on remarquera que la colonne des opérations indique de suite quatre *multiplications*, deux *soustractions* et une *division*. Ainsi l'on pourra, au besoin, n'employer que trois cartons des opérations; pour cela il suffira d'introduire dans la machine un appareil qui, par exemple, après la première multiplication, retienne le carton relatif à cette opération, et ne lui permette d'avancer, pour être remplacé par / un autre, que lorsque cette même opération aura été répétée quatre fois. Dans l'exemple précédent, nous avons vu que, pour trouver la valeur de x, il fallait commencer par écrire les coefficients m, n, d, m', n', d' sur huit colonnes, en répétant ainsi deux fois n et n'; d'après la même méthode, si l'on voulaie également calculer y, on aurait dû écrire ces mêmes coefficient sur douze colonnes différentes. Mais il est possible de simplifier cette opération et d'éviter les chances d'erreurs, qui augmentent à mesure que la quantité des nombres à écrire, avant de faire opérer la machine, devient plus grand. Pour cela souvenons-nous que tout nombre écrit sur une colonne, afin d'être combiné arithmétiquement avec un autre nombre, doit être effacé de la colonne sur laquelle il se trouve et transporté dans le *moulin*. Or, dans l'exemple que nous avons discuté, prenons les deux coefficients m et n', qui chacun doivent entrer dans deux produits différents, c'est-à-dire m dans $m\,n'$ et md', n' dans mn' et $n'd$. Ces coefficients seront écrits sur les colonnes V_0 et V_4. Si l'on commence la série des opérations par le produit de m par n', on effacera ces nombres sur les colonnes V_0 et V_4 pour les transporter dans le moulin qui les multipliera ensemble, puis ordonnera à la machine d'écrire le résultat sur la colonne V_6 par exemple. Mais comme ces nombres doiventencore servir chacun à une opération, il faudra qu'ils se trouvent de nouveau écrits quelque part; pour cela, en même temps que le moulin effectue leur produit, la machine les écrira de nouveau sur deux colonnes qui lui seront indiquées par les cartons, et comme dans le cas actuel rien ne s'oppose à ce qu'ils reprennent leurs premières places, on les supposera inscrits sur V_0 et V_4, d'où ils ne disparaîtront enfin, pour ne plus être reproduits, que lorsqu'ils auront subi les autres combinaisons auxquelles ils doivent être soumis.

On voit donc que l'ensemble des opérations nécessaires pour résoudre les deux équations du premier degré en question, pourra en définitive être représenté par le tableau suivant: /

Afin de diminuer autant que possible les chances d'erreur dans l'écriture des données numériques du problème, on les écrit successivement sur une des colonnes du moulin; puis, au moyen de cartons disposés à cet effet, ces mêmes nombres vont se placer sur les colonnes convenables, sans que l'opérateur ain aucunement à s'en inquiéter; de cette manière toute son attention se reportera sur la simple écriture de ces mêmes nombres.

D'après ce qui vient d'être exposé, on voit que l'ensemble des colonnes des variables peut être considéré comme un *magasin* de nombres qui y sont accumulés par le moulin et qui, / obéissant aux ordres transmis à la machine par le moyen des cartons, passent alternativement du moulin au magasin et du magasin au moulin, pour y

Colonnes sur lesquelles sont écrites les données primitives du problème	*Nombre des opérations*	*Cartons des opérations*		*Cartons des variables*			Marche des opérations
		Nombre des cartons des opérations	*Sigues indiquant la nature des opér*	*Colonnes soumises des opérations*	*Colonnes qui reçoivent le résultat des opérations*	*Indication des nouvelles colonnes sur lesquelles sont écrites les variables*	
$V_0 = m$	1	1	×	$V_0 \times V_4 =$	V_6.......	$\begin{cases} V_0 \text{ sur } V_0 \\ V_4 \textit{ id. } V_4 \end{cases}$	$V_6 = mn'$
$V_1 = n$	2	„	×	$V_3 \times V_1 =$	V_7.......	$\begin{cases} V_1 \textit{ id. } V_1 \\ V_3 \textit{ id. } V_3 \end{cases}$	$V_7 = m'n$
$V_2 = d$	3	„	×	$V_2 \times V_4 =$	V_8........	$\begin{cases} V_2 \textit{ id. } V_2 \\ V_4 \textit{ id.} \ldots \end{cases}$	$V_8 = dn'$
$V_3 = m'$	4	„	×	$V_5 \times V_1 =$	V_9.......	$\begin{cases} V_1 \textit{ id.} \ldots \\ V_5 \textit{ id. } V_5 \end{cases}$	$V_9 = d'n$
$V_4 = n'$	5	„	×	$V_0 \times V_5 =$	V_{10}.......	$\begin{cases} V_0 \textit{ id.} \ldots \\ V_5 \textit{ id.} \ldots \end{cases}$	$V_{10} = d'm$
$V_5 = d'$	6	„	×	$V_2 \times V_3 =$	V_{11}.......	$\begin{cases} V_2 \textit{ id.} \ldots \\ V_3 \textit{ id.} \ldots \end{cases}$	$V_{11} = dm'$
	7	2	−	$V_0 \times V_7 =$	V_{12}.......	$\begin{cases} V_6 \textit{ id.} \ldots \\ V_7 \textit{ id.} \ldots \end{cases}$	$V_{12} = mn' - m'n$
	8	„	−	$V_8 - V_9 =$	V_{13}.......	$\begin{cases} V_8 \textit{ id.} \ldots \\ V_9 \textit{ id.} \ldots \end{cases}$	$V_{13} = dn' - d'n$
	9	„	−	$V_{10} - V_{11} =$	V_{14}.......	$\begin{cases} V_{10} \textit{ id.} \ldots \\ V_{11} \textit{ id.} \ldots \end{cases}$	$V_{14} = d'm - dm'$
	10	3	÷	$\dfrac{V_{13}}{V_{12}} =$	V_{15}.......	$\begin{cases} V_{13} \textit{ id.} \ldots \\ V_{12} \textit{ id } V_{12} \end{cases}$	$V_{15} = \dfrac{dn' - d'n}{mn' - m'n} = x$
	11	„	÷	$\dfrac{V_{14}}{V_{12}} =$	V_{16}.......	$\begin{cases} V_{12} \textit{ id.} \ldots \\ V_{14} \textit{ id.} \ldots \end{cases}$	$V_{16} = \dfrac{d'm - dm'}{mn' - m'n} = y$

subir les transformations requises par la nature du calcul à exécuter.

Jusqu'ici, il n'a point encore été parlé des signes des résultats, et la machine serait loin d'être parfaite si elle n'était pas capable d'exprimer et de combiner entre elles les quantités positives et négatives. Pour arriver à ce but, au-dessus de chaque colonne, tant du moulin que du magasin, se trouve un disque semblable à ceux qui composent les colonnes. Selon que le chiffre correspondant au nombre écrit sur la pile inférieure est pair ou impair, ce dernier sera considéré comme positif ou négatif. Cela posé, voici une manière de concevoir comment les signes peuvent se combiner algébriquement dans la machine. Lorsqu'un nombre devra être transporté du magasin dans le moulin et *vice versâ*, il le sera toujours avec son signe, ce qui se fera par l'intermédiaire des cartons, ainsi qu'il a été dit précédemment. Soient donc deux nombres placés avec leurs signes respectifs dans le moulin et sur lesquels on doive opérer arithmétiquement. Supposons qu'il s'agisse d'abord de les sommer ensemble: les cartons des opérations commanderont l'addition; si les deux nombres ont le même signe, l'un des deux s'effacera entièrement sur la colonne où il était écrit et ira s'ajouter sur la colonne qui continet l'autre nombre; la machine, au moyen d'un appareil, pourra empêcher pendant cette opération tout mouvement dans le disque des signes qui appartient à la colonne sur laquelle se fait l'addition, et alors le résultat restera avec le signe qu'avaient les deux nombres donnés. Lorsque les deux nombres ont deux signes différents, l'addition qui était ordonnée par le carton se change en soustraction par l'intermédiaire des mécanismes qui sont mis en jeu à cause de cette différence de signe. Comme la soustraction ne peut s'opérer que sur le plus grand des deux nombres, si on fait en sorte que le disque des signes du plus grand nombre n'ait aucun mouvement / pendant que le plus petit de ces nombres s'efface de desus sa colonne pour être soustrait de l'autre, il restera un résultat qui aura le signe de ce dernier, ainsi que cela doit être. Les combinaisons auxquelles done lieu la soustraction algébrique, sont analogues aux précédentes. Passons à la multiplication. Quand les deux nombres à multiplier ont le même signe, le résultat est positif; si les deux signes sont différents, le produit doit être négatif. Pour que la machine agisse conformément à cette loi, il suffira de concevoir que, sur la colonne qui contient le produit des deux nombres donnés, les chiffres indiquant leurs signes se soient ajoutés sur le disque des signes du produit; alors il est clair que si les chiffres des signes sont ou tous les deux pairs, ou tous les deux impairs, leur somme sera un nombre pair, et par conséquent exprimera une quantité positive; si, au contraire, les deux chiffres des signes sont l'un pair et l'autre impair, leur somme sera un nombre impair, et par conséquent exprimera une quantité négative. Pour la division, au lieu d'ajouter les chiffres des disques, il faudra les soustraire l'un de l'autre, ce qui donnera lieu à des résultats analogues aux précédents, c'est-à-dire que, si ces chiffres sont tous les deux pairs ou impairs, le reste de cette soustraction sera pair; il sera impair dans le cas contraire. Lorsque je parle d'ajouter ou de soustraine l'un de l'autre les nombres exprimés par les chiffres des signes, je n'entends que faire avancer ou rétrograder l'un des disques d'un nombre de divisions égal à celui qui est exprimé par le chiffre de l'autre disque. On voit donc, d'après ce qui vient d'être exposé, qu'il est possible de combiner mécaniquement les signes des quantités, de manière à obtenir des résultats conformes à ceux qu'indique l'algèbre.[a] /

La machine est non-seulement capable d'exécuter les calculs numériques qui dépendent d'une formule algébrique donnée, mais elle est encore propre aux calculs analytiques dans lesquels on a une ou plusieurs variables à considérer. Pour cela il faut admettre que l'expression analytique sur laquelle on doit opérer soit développée suivant les puissances de la variable, ou suivant des fonctions déterminées de cette même variable, comme seraient, par exemple, les fonctions circulaires; il en doit être de même du résultat auquel on veut arriver. Or, si l'on suppose qu'au-dessus de chaque colonne du magasin on ait écrit les puissances ou les fonctions de la variable ordonnées selon la loi du développement que l'on considère, les coefficients de ces différents termes pourront être écrits sur les colonnes inférieures qui leur correspondent. On aura, de cette manière, représenté un développement analytique. Maintenant, en supposant que la position des différents termes qui le composent soit invariable, le problème se réduira à calculer leurs coefficients selon les lois voulues par la nature de la question. Pour mieux faire comprendre la chose, prenons l'exemple suivant, bien simple, dans lequel il s'agit de multiplier $(a + bx^1)$ par $(A + B\cos^1 x)$. Nous commencerons par écrire x^0, x^1, $\cos^0 x$, $\cos^1 x$ au-dessus des colonnes V_0, V_1, V_2 et V_3; puis comme, d'après la forme des deux fonctions à combiner, les termes qui doivent composer le produit sont de la nature suivante, $x^0 \cdot \cos^0 x$, $x \cdot \cos^1 x$, x^1, $\cos^0 x$, $x^1 \cos^1 x$, on les écrira au-dessus des colonnes V_4, V_5, V_6 et V_7. Les coefficients de x^0, x^1, $\cos^0 x$, $\cos^1 x$ étant donnés, par le moyen du moulin on les fera passer sur les colonnes V_0, V_1, V_2 et V_3. Telles sont les données primitives du problème; c'est maintenant à la machine qu'il appartient d'en opérer la solution, c'est-à-dire de trouver les coefficients

[a] N'ayant point eu le loisir de discuter avec Mr Babbage la manière d'introduire dans sa machine la combinaison des signes algébriques, je n'ai point la prétention d'exposer ici la méthode dont il se sert à cette fin; mais j'ai cru devoir y suppléer, pensant que cet écrit eût été imparfait si je n'avais indiqué un des moyens que l'on peut employer pour résoundre ce point essentiel du problème en question.

qui doivent être inscrits sur V_4, V_5, V_6, V_7. Dans ce but, comme la loi de formation de ces mêmes coefficients est connue, la loi de formation de ces mêmes coefficients est connue, la machine agira par l'intermédiaire des cartons, de la manière indiquée dans le tableau suivant: /

On doit concevoir maintenant que le principe développé dans l'exemple précédent puisse s'appliquer en général à toute espèce d'opération que l'on pourrait se proposer sur les séries soumises au calcul. Il suffira de connaître la loi de formation des différents coefficients, d'inscrire cette loi dans les cartons de la machine, qui alors exécutera elle-même tous les calculs convenables pour arriver au résultat proposé. S'il s'agissait, par exemple, d'une série récurrente, la loi de formation des différents termes étant la même pour tous, les mêmes opérations qui se feront pour l'un d'eux se répéteront pour les autres; seulement il y aura changement de lieu de l'opération, c'est-à-dire qu'elle se fera sur des colonnes différentes. Comme généralement toute expression analytique est susceptible d'être exprimée par une série ordonnée selon certaines fonctions de la variable, on voit que la machine embrassera tous les calculs analytiques qui se réduisent en définitive à la formation des coefficients suivant certaines lois, et à leur distribution par rapport aux variables.

De ce qui vient d'être exposé l'on peut déduire cette conséquence importante, c'est que, puisque les cartons indiquent seulement la nature des opérations à faire et les colonnes des variables sur lesquelles elles doivent être exécutées, ces cartons auront eux-mêmes toute la généralité d'analyse, dont ils ne sont qu'une simple traduction. Examinons encore quelques-unes des difficultés que la machine doit surmonter pour que son assimilation à l'analyse soit plus complète. Il y a certaines fonctions qui doivent changer de nature lorsqu'elles passent par zéro ou l'infini, ou dont les valeurs ne sauraient être admises au delà de ces limites. Lorsque ces cas se présentent, la machine au moyen d'une sonnerie peut avertir que le passage par zéro ou l'infini a lieu, et alors elle s'arrête, jusqu'à ce que le gardien l'ait remise en action pour telle autre opération qu'on voudra lui ordonner. Si cette opération est prévue, la machine, au lieu de sonner, se disposera de manière à présenter les nouveaux cartons / tons relatifs à l'opération qui doit succéder au passage par zéro et l'infini. Ces nouveaux cartons peuvent faire suite aux premiers et n'entrer en jeu que lorsque les deux circonstances indiquées précédemment auront lieu. Considérons un terme de la forme ab^n; puisque les cartons ne sont qu'une traduction de la formula analytique, il faudra, dans le cas actuel, que leur nombre soit

Colonnes au-dessus desquelles sont écrites les fonetions de la variable		*Coefficients*		*Cartons des opérations*		*Cartons des variables*			
		Donués	*A former*	*Nombre des opérations*	*Nature de l'opération*	*Colonnes soumises aux opérations*	*Colonnes sur lesquelles sont ecrits les resultats des operations*	*Indication des nouvelles colonnes sur lesquelles sont ecrits les coefficients soumis aux opérations*	*Résultats des opérations*
x^0	V_0	a							
x'	V_0	b							
$\cos^0 x$	V_2	A							
$\cos^1 x$	V_3	B							
$x^0 \cos^0 x$	V_4		aA	1	$\times$	$V_0 \times V_2 =$	V_4	$\begin{cases} V_0 \text{ sur } V_0 \\ V_2 \; id. \; V_2 \end{cases}$	$V_4 = aA$ coefficients de $x^0 \cos^0 x$
$x^0 \cos^1 x$	V_5		aB	2	$\times$	$V_0 \times V_3 =$	V_5	$\begin{cases} V_0 \; id. \ldots \\ V_3 \; id. \; V_3 \end{cases}$	$V_5 = aB$ *id.* $x^0 \cos^1 x$
$x' \cos^0 x$	V_6		bA	3	$\times$	$V_1 \times V_2 =$	V_6	$\begin{cases} V_1 \; id. \; V_1 \\ V_2 \; id. \ldots \end{cases}$	$V_6 = bA$ *id.* $x' \cos^0 x$
$x' \cos^1 x$	V_7		bB	4	$\times$	$V_1 \times V_3 =$	V_7	$\begin{cases} V_1 \; id. \ldots \\ V_3 \; id. \ldots \end{cases}$	$V_7 = bB$ *id.* $x' \cos^1 x$

le même, quelle que soit la valeur de n, c'est-à-dire quel que soit le nombre des multiplications à faire pour élever b à la puissance $n^{\text{ième}}$ (nous supposons pour le moment que n soit un nombre entier). Or, comme l'exposant n indique que b doit être multiplié n fois par lui-même, toutes ces opérations étant de la même nature, il suffira d'employer pour cela un seul carton des opérations, celui qui ordonne la multiplication.

Mais lorsque n sera donné pour le cas particulier que l'on veut calculer, il sera encore nécessaire que la machine borne ses multiplications aux nombres convenables. Or, voici comment l'opération peut être disposée. Les trois nombres a, b et n seront écrits sur autant de colonnes distinctes du magasin, que nous désignerons par V_0, V_1, V_2; le résultat ab^2 viendra se transcrire sur la colonne V_3. Lorsque le nombre n aura été introduit dans la machine, un carton ordonnera à un *compteur* de marquer $(n - 1)$ et en même temps fera exécuter la multiplication de b par b. Quand celle-ci sera achevée, le compteur aura effacé une unité et ne marquera plus que $(n - 2)$, et la machine ordonnera de nouveau au nombre b écrit sur la colonne V_1 de se multiplier avec le produit b^2 écrit sur la colonne V_3, ce qui donnera b^3. Alors une nouvelle unité s'effacera sur le compteur, et les opérations indiquées se répéteront jusqu'à ce que le compteur ne marque plus que zéro. Ainsi le nombre b^n se trouvera écrit sur la colonne V_3, et la machine poursuivant le cours de ses autres opérations, ordonnera le produit de b^n par a, et la calcul demandé se trouvera achievé sans que le nombre de cartons à employer ait dû varier avec la valeur de n. Si n hait négatif, les cartons, au lieu d'ordonner la multiplication / de a par b^n, ordonneraient la division, ce qui peut se concevoir, puisque tout nombre peut être écrit avec son signe respectif et réagir de cette manière sur la nature des opérations à exécuter. Enfin si n était fractionnaire et de la forme p/q, on emploierait une colonne de plus pour écrire q, et la machine ferait agir deux appareils, l'un pour élever b à la puissance n, l'autre pour extraire la racine $q^{\text{ième}}$ du nombre ainsi obtenu.

On peut demander, par exemple, de multiplier une expression de la forme $ax^m + bx^n$ par une autre $Ax^p + Bx^q$, puis de réduire le produit aux moindres termes lorsque leurs indices sont égaux. Les deux facteurs étant ordonnés par rapport à x, le résultat général de la multiplication serait

$$Aax^{m+p} + Abx^{p+n} + aBx^{m+q} + bBx^{n+q}.$$

Jusqu'ici l'opération ne présente pas de difficultés; mais supposons que l'on ait $m = p$ et $n = q$, et que l'on veuille réduire les deux termes moyens à un seul $(Ab + Ba)\ x^{m+q}$. Pour cela les cartons pourront ordonner que $m + q$ et $n + p$ soient transportés dans le moulin pour y être soustraits l'un de l'autre; si le reste est nul, comme cela doit avoir lieu dans l'hypothèse admise, le moulin ordonnera à d'autres cartons de lui porter les coefficients Ab et aB pour les sommer ensemble, et les donner ainsi pour coefficient au terme unique $x^{n+p} = x^{m+q}$.

Cet exemple nous montre comment les cartons peuvent reproduire toutes les opérations qu'exécute l'intelligence pour arriver à un résultat déterminé, lorsque ces opérations sont elles-mêmes susceptibles d'être nettement précisées.

Examinons encore l'expression suivante:

$$2 \cdot \frac{2^2 \cdot 4^2 \cdot 6^2 \cdot 8^2 \cdot 10^2 \cdot \ldots \cdot (2n)^2}{1^2 \cdot 3^2 \cdot 5^2 \cdot 7^2 \cdot 9^2 \cdot \ldots \cdot (2n-1)^2 \cdot (2n+1)^2}$$

que l'on sait devenir égale au rapport de la circonférence au diamètre, lorsque n est infini. On peut demander à la machine non-seulement d'exécuter le calcul de cette expression fractionnaire, mais encore d'indiquer immédiatement que sa valeur devient / celle même du rapport de la circonférence au diamètre lorsque n est infini, cas dans lequel le calcul serait impossible. Observons que l'on exigerait ainsi que la machine interprétât un résultat non évident par lui-même, ce qui n'est point dans ses attributions, puisqu'elle n'est point un être pensant. Cependant lorsque le cos. de $n = \infty$ a été prévu,[a] un carton peut ordonner immédiatement la substitution de la valeur de π (π étant le rapport de la circonférence au diamètre), sans passer par la série des calculs indiqués. Il suffirait pour cela qu'il y eût dans la machine un carton spécial destiné à former immédiatement le nombre π sur la colonne qui lui serait indiquée. C'est ici le cas de parler d'une troisième espèce de cartons que l'on peut appeler *cartons des nombres*. Il y a certains nombres, tels que celui qui exprime le rapport de la circonférence au diamètre, les nombres de Bernoulli, etc., qui se présentent fréquemment dans les calculs. Pour n'être pas obligé de les calculer chaque fois que l'on doit les employer, on peut combiner certains cartons destinés à les donner tout faits dans le moulin, d'où ils vont ensuite se placer dans les colonnes du magasin qui leur sont destinées. Par ce moyen, la machine sera encore susceptible des

[a] *Sic*; this is a printer's error in the original, in which "le cos. de $n = \infty$" should have read "le cas de $n = \infty$".

simplifications que présente l'emploi des tables numériques. Il serait également possible d'introduire, au moyen de ces cartons, les logarithmes des nombres; mais peut-être ici ne serait-ce pas la méthode la plus convenable et la plus prompte, car la machine pourrait se prêter à d'autres combinaisons plus expéditives, fondées sur la rapidité avec laquelle s'exécutent les quatre premières opérations de l'arithmétique. Pour donner une idée de cette rapidité, il suffira de dire que Mr Babbage pense pouvoir, par le moyen de sa machine, faire en *trois minutes* le produit de deux nombres composés de vingt chiffres chacun.

Peut-être le nombre immense de cartons qu'exigerait la solution d'un problème un peu compliqué, pourrait-il paraître un obstacle. Mais il ne semble pas devoir en être ainsi: le nombre / de cartons que l'on peut employer n'a pas de limites. Il y a certaines étoffes qui pour être confectionnées n'exigent pas moins de *vingt mille* cartons, et cette quantité peut certainement être de beaucoup dépassée.

En résumant ce qui vient d'être dit sur la machine analytique, on peut conclure qu'elle est fondée sur deux principes: le premier, qui consiste en ce que tout calcul arithmétique dépend en définitive de quatre opérations principales, l'addition, la soustraction, la multiplication et la division; le second, en ce que tout calcul analytique peut se réduire à calculer les coefficients des différents termes d'une série. Si ce dernier principle est vrai, toutes les opérations de l'analyse sont du domaine de la machine. D'un autre côté, l'emploi des cartons offre une généralité égale à celle des formules algébriques, car, de même qu'une formule de ce genre indique simplement la nature et l'ordre des opérations à faire pour arriver à un résultat déterminé, les cartons ne font que commander à la machine ces mêmes opérations; mais pour que les mécanismes puissent agir, il faut encore introduire dans chaque cas particulier les données numériques du problème. Ainsi une même série de cartons servira pour toutes les questions d'une même nature, dans lesquelles il n'y aura que les données numériques de changées. De cette manière les cartons ne sont qu'une traduction des formules algébriques, ou pour mieux dire, une autre forme d'écriture analytique.

Comme la machine a un mode d'agir qui lui est propre, dans chaque cas particulier il sera nécessaire de disposer la série des calculs conformément aux moyens qu'elle possède; car tel procédé qui serait extrêmement facile pour un calculateur, deviendrait long et compliqué pour la machine et *vice versâ*.

Considéré sous le point de vue le plus général, l'objet essential de la machine étant de calculer, d'après les lois qui lui sont dictées, la valeur de coefficients numériques qu'elle doit ensuite distribuer convenablement sur les colonnes qui représentent / les variables, il s'ensuit que l'interprétation des formules et des résultats est en dehors de ses attributions, à moins toutefois que cette interprétation ne soit elle-même susceptible d'être exprimée par le moyen des symboles dont elle fait usage. Ainsi elle n'est point elle-même l'être qui pense, mais on peut la considérer comme l'être qui exécute les conceptions de l'intelligence. Les cartons reçoivent l'empreinte de ces conceptions, et transmettent aux différents mécanismes dont se compose la machine les ordres nécessaires pour agir. Une fois que la machine sera construite, la difficulté se reportera donc sur la confection des cartons; mais comme ceux-ci ne sont que la traduction de formules algébriques, par le moyen de simples notations il sera facile d'en confier l'exécution à un ouvrier. Ainsi tout le travail intellectuel se bornera à la préparation des formules, qui devront être aptes à être calculées par la machine.

Maintenant, en admettant qu'une telle machine soit exécutable, on pourra demander quelle en sera l'utilité? En résumé, elle présentera les avantages suivants: 1° Exactitude; on sait que les calculs numériques sont généralement l'écueil de la solution des problèmes, car les erreurs s'y glissent aisément, et il n'est pas toujours facile de les reconnaître. Or, la machine, par la nature même de son mode d'agir, qui n'exige aucunement l'intervention de la main de l'homme durant le cours de ses opérations, présente toute espèce de garantie au sujet de l'exactitude; d'ailleurs, elle porte son contrôle avec elle-même, car à la fin de chaque opération elle donne imprimés, non-seulement le résultat, mais encore les données numériques de la question, de sorte qu'il est aisé de vérifier si cette question a été posée avec exactitude. 2° Économie de temps: pour s'en convaincre, il suffira de rappeler qu'une multiplication de deux nombres composés chacun de 20 chiffres demande tout au plus trois minutes. D'ailleurs, lorsque l'on devra faire une longue série de calculs identiques, comme ceux qu'exige la formation de tables numériques, on pourra mettre en jeu la machine / de manière à donner plusieurs résultats à la fois, ce qui abrégera de beaucoup l'ensemble des opérations. 3° Economie d'intelligence: un simple calcul d'arithmétique exige le concours d'une personne ayant queique capacité; lorsqu'on passe à des calculs plus compliqués, et qu'on veut faire usage de formules algébriques dans des cas particuliers, il faut déjà posséder des connaissances qui supposent des études mathématiques préliminaires de quelque étendue. Or, la machine, pouvant faire elle-même toutes ces opérations purement matérielles, épargne le travail d'intelligences qui peuvent être employées plus utilement. Ainsi la machine pourra être considérée comme une vraie manufacture de chiffres, qui prêtera son secours aux sciences et aux arts utiles qui s'appuient sur les nombres. Or, qui pourrait prévoir les conséquences d'une telle invention? En effet, combien d'observations précieuses restent inutiles aux progrès des sciences, parce qu'il n'y a pas de forces suffisantes pour en calculer les résultats! Que de découragement la perspective d'un long et aride calcul ne jette-t-elle pas dans l'âme de l'homme de génie qui ne demande que du temps pour méditer et qui se le voit ravi par le matériel des opérations! Et pourtant c'est par la voie laborieuse de l'analyse qu'il doit arriver à la vérité; mais il ne peut la suivre sans être guidé par des nombres, car sans les nombres il n'est pas donné de pouvoir soulever le voile qui couvre les mystères de la nature. Ainsi la pensée de former un instrument capable d'aider la faiblesse de l'homme dans de telles recherches, est une conception qui, venant à se réaliser, marquerait une époque glorieuse dans l'histoire des sciences. Toutes les pièces, tous les rouages qui composent cet immense appareil ont été combinés, leur action a été étudiée, mais ils n'ont pu être encore assemblés. La confiance que doit inspirer le génie de Mr Babbage rend légitime l'espoir que cette entreprise sera couronnée de succès; en rendant hommage à l'intelligence qui la dirige, faisons des œuvre s'accomplisse. /

VI

ADDITION TO THE MEMOIR OF M. MENABREA ON THE ANALYTICAL ENGINE[a]

From *Philosophical Magazine*, Vol. 23 (1843), pp. 235–9

Much misapprehension having arisen as to the circumstances attending the invention and construction of Mr Babbage's calculating engines, it is necessary to state *from authority* the facts relating to them.

In 1823, Mr Babbage, who had previously invented an engine for calculating and printing tables by means of *differences*, undertook, at the desire of the Government, to superintend the construction of such an engine. He bestowed his whole time upon the subject for many years, refusing for that purpose other avocations which would have been attended with considerable pecuniary advantage. During this period about £17,000 had been expended by the Government in the construction of the Difference Engine. A considerable part of this

---

[a] This article is a brief history of Babbage's calculating engine and his dealings with the Government. It appeared in the *Philosophical Magazine* in September, 1843, and was written to accompany Ada Lovelace's 'Sketch of the Analytical Engine' which appeared in Taylor's *Scientific Memoirs* at about the same time (and which follows immediately in this volume of the *Works of Babbage*). A note by the editor of *Philosophical Magazine*, which was also published by Taylor, stated:

> The rules prescribed for the publication of the Foreign Scientific Memoirs prevented the Editor from inserting in that work the following authorized statement of the facts connected with the history of Mr Babbage's calculating engines. As those facts may be interesting to our readers, we take this opportunity of communicating them.

The article was published anonymously, but was written by Babbage. It was reprinted by H. P. Babbage, with the title 'Statement of the circumstances attending the invention and construction of Mr. Babbage's calculating engines', in *Babbage's calculating engines* (London, 1889), pp. 2–4.

sum had from time to time been advanced by Mr Babbage for the payment of the workmen, and was of course repaid; but it was never contemplated by either party that any portion of this sum should be appropriated to Mr Babbage himself, and in truth not one single shilling of the money was in any shape whatever received by Mr Babbage for his invention, his time, or his services, a fact which Sir Robert Peel admitted in the House of Commons in March, 1843.

Early in 1833 the construction of this engine was suspended on account of some dissatisfaction with the workmen, which it is now unnecessary to detail. It was expected that the interruption, which arose from circumstances over which Mr Babbage had no control, would be only temporary. About twelve months after the progress of the Difference Engine had been thus suspended, Mr Babbage discovered a principle of an entirely new order, the power of which over the most complicated arithmetical operations seemed nearly unbounded. The invention / of simpler mechanical means for executing the elementary operations of that engine, now acquired far greater importance than it had hitherto possessed.

In the engine for calculating by differences, such simplifications affected only about a hundred and twenty similar parts, while in the new, or Analytical Engine, they might affect several thousand. The Difference Engine might be constructed with more or less advantage, by employing various mechanical modes for the operation of addition. The Analytical Engine could not exist without inventing for it a method of mechanical addition possessed of the utmost simplicity. In fact it was not until upwards of twenty different modes for performing the operation of addition had been designed and drawn, that the necessary degree of simplicity required for the Analytical Engine was ultimately attained.

These new views acquired great additional importance from their bearings upon the Difference Engine already partly executed for the Government; for if such simplifications should be discovered, it might happen that the Analytical Engine would execute with greater rapidity the calculations for which the Difference Engine was intended; or that the Difference Engine would itself be superseded by a far simpler mode of construction.

Though these views might, perhaps, at that period, have appeared visionary, they have subsequently been completely realized.

To have allowed the construction of the Difference Engine to be resumed while these new views were withheld from the Government, would have been improper; yet the state of uncertainty in which those views were then necessarily involved, rendered any written communication respecting their probable bearing on that engine a matter of very great difficulty. It therefore appeared to Mr Babbage that the most straightforward course was to ask for an interview with the head of the Government, and to communicate to him the exact state of the case. Various circumstances occurred to delay, and ultimately to prevent that interview.

From the year 1833 to the close of 1842, Mr Babbage repeatedly applied to the Government for its decision upon the subject. These applications were unavailing. Years of delay and anxiety followed each other, impairing those energies which were now directed to the invention of the Analytical Engine. This state of uncertainty had many injurious effects. It prevented Mr Babbage from entering into any engagement with other governments respecting the Analytical Engine, by which he might have been enabled to employ a greater number of assistants, and thus to have applied his faculties only to the highest departments of the subject, instead of exhausting them on inferior objects, that might have been executed with less fatigue by other / heads. It also became necessary, from motives of prudence, that the heavy expense incurred for this purpose should be spread over a period of many years. This consideration naturally caused a new source of anxiety and risk, arising from the uncertain tenure of human life and of human faculties, a reflection ever present to distract and torment the mind, and itself calculated to cause the fulfilment of its own forebodings.

Amidst such distractions the author of the Analytical Engine has steadily pursued his single purpose. The numberless misrepresentations of the facts connected with both engines have not induced him to withdraw his attention from the new invention; and the circumstance of his not having printed a description of either engine has arisen entirely from his determination never to employ his mind upon the *description* of those machines so long as a single difficulty remained which might *limit* the *power* of the Analytical Engine. The drawings, however, and the notations have been freely shown; and the great principles on which the Analytical Engine is founded have been explained and discussed with some of the first philosophers of the present day. Copies of the engravings were sent to the libraries of several public institutions, and the effect of the publicity thus given to the subject is fully proved by its having enabled a distinguished

Italian geometer to draw up from these sources an excellent account of that engine.*

Throughout the whole of these labours connected with the Analytical Engine, neither the science, nor the institutions, nor the Government of his country have ever afforded him the slightest encouragement. When the invention was noticed in the House of Commons, one single voice† alone was raised in its favour.

During nearly the whole of a period of upwards of twenty years, Mr Babbage had maintained, in his own house, and at his own expense, an establishment for aiding him in carrying out his views, and in making experiments, which most materially assisted in improving the Difference Engine. When that work was suspended he still continued his own enquiries, and having discovered principles of far wider extent, he ultimately embodied them in the Analytical Engine.

The establishment necessary in the former part of this period for the actual construction of the Difference Engine, and of the extensive drawings which it demanded, as well as for the formation of those tools which were contrived to overcome the novel / difficulties of the case, and in the latter part of the same period by the drawings and notations of the Analytical Engine, and the experiments relating to its construction, gave occupation to a considerable number of workmen of the greatest skill. During the many years in which this work proceeded, the workmen were continually changing, who carried into the various workshops in which they were afterwards employed the practical knowledge acquired in the construction of these machines.

To render the drawings of the Difference Engine intelligible, Mr Babbage had invented a compact and comprehensive language (the Mechanical Notation), by which every contemporaneous or successive movement of this machine became known. Another addition to mechanical science was subsequently made in establishing principles for the *lettering* of drawings; one consequence of which is, that although many parts of a machine may be projected upon any plan, it will be easily seen, by the nature of the letter attached to each working point, to which of those parts it really belongs.

* Of M. Menabrea's treatise, which appeared in the Bibliothèque Universelle de Genève for October last, a translation is given in the 12th Part of the Scientific Memoirs just published, with copious and valuable explanatory notes by the translator.

† That of Mr Hawes, Member for Lambeth.

By the means of this system, combined with the Mechanical Notations, it is now possible to express the forms and actions of the most complicated machine in language which is at once condensed, precise, and universal.

At length, in November, 1842, Mr Babbage received a letter from the Chancellor of the Exchequer, stating that Sir Robert Peel and himself had jointly and reluctantly come to the conclusion that it was the duty of the Government, on the ground of expense, to abandon the further construction of the Difference Engine. The same letter contained a proposal to Mr Babbage, on the part of government, that he should accept the whole of the drawings, together with the part of the engine already completed, as well as the materials in a state of preparation. This proposition he declined.

The object of the Analytical Engine (the drawings and the experiments for which have been wholly carried on at Mr Babbage's expense, by his own draftsmen, workmen and assistants) is to convert into numbers all the formulae of analysis, and to work out the algebraical development of all formulae whose laws are known.

The present state of the Analytical Engine is as follows:

All the great principles on which the discovery rests have been explained, and drawings of mechanical structures have been made, by which each may be carried into operation.

Simpler mechanisms, as well as more extensive principles than were required for the Difference Engine, have been discovered for all the elementary portions of the Analytical Engine, and numerous drawings of these successive simplifications exist. /

The mode of combining the various sections of which the engine is formed has been examined with unceasing anxiety, for the purpose of reducing the whole combination to the greatest possible simplicity. Drawings of almost all the plans thus discussed have been made, and the latest of the drawings (bearing the number 28) shows how many have been superseded, and also, from its extreme comparative simplicity, that little further advance can be expected in that direction.

Mechanical Notations have been made both of the actions of detached parts and of the general action of the whole, which cover about four or five hundred large folio sheets of paper.

The original rough sketches are contained in about five volumes.

There are upwards of one hundred large drawings.

No part of the construction of the Analytical Engine has yet been commenced. A long series of experiments have, however, been made upon the art of shaping metals; and the tools to be employed for that purpose have been discussed, and many drawings of them prepared. The great object of these enquiries and experiments is, on the one hand, by simplifying as much as possible the construction, and on the other, by contriving new and cheaper means of execution, at length to reduce the expense within those limits which a private individual may command. /

VII

SKETCH OF THE ANALYTICAL ENGINE[a]

BY A. A. LOVELACE

From *Scientific Memoirs*, Vol. 3 (1843), pp. 666–731

[From the *Bibliothèque Universelle de Genève*, No. 82. October, 1842.[b]]

[Before submitting to our readers the translation of M. Menabrea's memoir 'On the Mathematical Principles of the Analytical Engine' invented by Mr Babbage, we shall present to them a list of the printed papers connected with the subject, and also of those relating to the Difference Engine by which it was preceded.

For information on Mr Babbage's '*Difference* Engine,' which is but slightly alluded to by M Menabrea, we refer the reader to the following sources:[c]

1. Letter to Sir Humphry Davy, Bart., P.R.S., on the Application of Machinery to Calculate and Print Mathematical Tables. By Charles Babbage, Esq., F.R.S. London, July, 1822. Reprinted, with a Report of the Council of the Royal Society, by order of the House of Commons, May, 1823.

2. On the Application of Machinery to the Calculation of Astronomical and Mathematical Tables. By Charles Babbage, Esq. Memoirs of the Astronomical Society, vol. i. part 2. London, 1822.

[a] The complete title of this work is usually given as 'Sketch of the Analytical Engine invented by Charles Babbage Esq. By L. F. Menabrea, of Turin, officer of the Military Engineers, with notes upon the memoir by the translator.'

[b] L. F. Menabrea 'Notions sur la machine analytique de M. Charles Babbage' (1842), reprinted in this volume.

[c] The sources listed here are reprinted in the *Works of Babbage* as follows: papers 1, 2, 3, 4, 5, 7, 8 and 10 as papers II, IV, VII, VIII, VI, IX, XI and XIII in Vol. 2, paper 6 as paper I in this volume, section C, and source 10 in Vol. 8, chapter XX.

3. Address to the Astronomical Society by Henry Thomas Colebrooke, Esq., F.R.S., President, on presenting the first Gold Medal of the Society to Charles Babbage, Esq., for the invention of the calculating engine. Memoirs of the Astronomical Society. London, 1822.

4. On the Determination of the General Term of a New Class of Infinite Series. By Charles Babbage, Esq. Transactions of the Cambridge Philosophical Society.

5. On Mr Babbage's New Machine for Calculating the Printing Mathematical Tables. Letter from Francis Baily, Esq., F.R.S., to M. Schumacher. No. 46, Astronomische Nachrichten. Reprinted in the Philosophical Magazine, May 1824.

6. On a Method of expressing by Signs the Action of Machinery. / By Charles Babbage, Esq. Philosophical Transactions, London, 1826.

7. On Errors common to many Tables of Logarithms. By Charles Babbage, Esq. Memoirs of the Astronomical Society, London, 1827.

8. Report of the Committee appointed by the Council of the Royal Society to consider the subject referred to in a communication received by them from the Treasury respecting Mr Babbage's Calculating Engine, and to report thereon, London, 1829.

9. Economy of Manufactures, chap. xx. octavo, London, 1832.

10. Article on Babbage's calculating engine. Edinburgh Review, July, 1834. No. 120. vol. lix.

The present state of the Difference Engine, which has always been the property of government, is as follows: The drawings are nearly finished, and the mechanical notation of the whole, recording every motion of which it is susceptible, is completed. A part of that engine, comprising sixteen figures, arranged in three orders of differences, has been put together, and has frequently been used during the last eight years. It performs its work with absolute precision. This portion of the Difference Engine, together with all the drawings, are at present deposited in the Museum of King's College, London.

Of the Analytical Engine, which forms the principal object of the present memoir, we are not aware that any notice has hitherto appeared, except a letter from the inventor to M. Quetelet, Secretary to the Royal Academy of Sciences at Brussels, by whom it was communicated to that body. We subjoin a translation of this letter, which was itself a translation of the original, and was not intended for publication by its author.

Royal Academy of Sciences at Brussels. General Meeting of the 7–8 of May, 1835

'A letter from Mr Babbage announces that he has for six months been engaged in making the drawings of a new calculating machine of far greater power than the first.

'I am myself astonished,' says Mr Babbage, 'at the power I have been enabled to give to this machine; a year ago I should not have believed this result possible. This machine is intended to contain a hundred variables (or numbers susceptible of changing); / each of these numbers may consist of twenty-five figures, $v_1, v_2, \ldots, v_n$ being any numbers whatever, n being less than a hundred; if $f(v_1, v_2, v_3, \ldots, v_n)$ be any given function which can be formed by addition, subtraction, multiplication, division, extraction of roots, or elevation to powers, the machine will calculate its numerical value; it will afterwards substitute this value in the place of v, or of any other variable, and will calculate this second function with respect to v. It will reduce to tables almost all equations of finite differences. Let us suppose that we have observed a thousand values of a, b, c, d, and that we wish to calculate them by the formula

$$p = \sqrt{\frac{a+b}{cd}}$$

the machine must be set to calculate the formula; the first series of the values of a, b, c, d must be adjusted to it; it will then calculate them, print them, and reduce them to zero; lastly, it will ring a bell to give notice that a new set of constants must be inserted. When there exists a relation between any number of successive coefficients of a series, provided it can be expressed as has already been said, the machine will calculate them and make their terms known in succession; and it may afterwards be disposed so as to find the value of the series for all the values of the variable.'

Mr Babbage announces, in conclusion, 'that the greatest difficulties of the invention have already been surmounted, and that the plans will be finished in a few months.'

In the Ninth Bridgewater Treatise, Mr Babbage has employed several arguments deduced from the Analytical Engine, which afford some idea of its powers. See Ninth Bridgewater Treatise, octavo, second edition. London, 1838.

Some of the numerous drawings of the Analytical Engine have been engraved on wooden blocks, and from these (by a mode contrived by Mr Babbage) various stereotype plates have been taken. They comprise

1. Plan of the figure wheels for one method of adding numbers.
2. Elevation of the wheels and axis of ditto.
3. Elevation of framing only of ditto.
4. Section of adding wheels and framing together.
5. Section of the adding wheels, sign wheels and framing complete.
6. Impression from the original wood block. /
7. Impressions from a stereotype cast of No. 6, with the letters and signs inserted. Nos. 2, 3, 4, and 5 were stereotypes taken from this.
8. Plan of adding wheels and of long and short pinions, by means of which *stepping* is accomplished.

N.B. This process performs the operation of multiplying or dividing a number by any power of ten.

9. Elevation of long pinions in the position for addition.
10. Elevation of long pinions in the position for stepping.
11. Plan of mechanism for carrying the tens (by anticipation), connected with long pinions.
12. Section of the chain of wires for anticipating carriage.
13. Sections of the elevation of parts of the preceding carriage.

All these were executed about five years ago. At a later period (August 1840) Mr Babbage caused one of his general plans (No. 25) of the whole Analytical Engine to be lithographed at Paris.

Although these illustrations have not been published,[a] on account of the time which would be required to describe them, and the rapid succession of improvements made subsequently, yet copies have been freely given to many of Mr Babbage's friends, and were in August, 1838, presented at Newcastle to the British Association for the Advancement of Science, and in August, 1840, to the Institute of France through M. Arago, as well as to the Royal Academy of Turin through M. Plana. *Editor*]

Those labours which belong to the various branches of the mathematical sciences, although on first consideration they seem to be the exclusive province of intellect, may, nevertheless, be divided into two

[a] The drawings were first published as '13 plates' and 'Plan No. 25 . . . of the Analytical Engine' by H. P. Babbage in *Babbage's calculating engines* (London, 1889). They are reproduced in the appendix of this volume of the *Works of Babbage*.

distinct sections; one of which may be called the mechanical, because it is subjected to precise and invariable laws, that are capable of being expressed by means of the operations of matter; whilst the other, demanding the intervention of reasoning, belongs more specially to the domain of the understanding. This admitted, we may propose to execute, by means of machinery, the mechanical branch of these labours, reserving for pure intellect that which depends on the reasoning faculties. Thus the rigid exactness of those laws which regulate numerical calculations must frequently have suggested the employment of material instruments, either for executing the whole / of such calculations or for abridging them; and thence have arisen several inventions having this object in view, but which have in general but partially attained it. For instance, the much admired machine of Pascal is now simply an object of curiosity, which, whilst it displays the powerful intellect of its inventor, is yet of little utility in itself. Its powers extended no further than the execution of the four* first operations of arithmetic, and indeed were in reality confined to that of the two first, since multiplication and division were the result of a series of additions and subtractions. The chief drawback hitherto on most of such machines is, that they require the continual intervention of a human agent to regulate their movements, and thence arises a source of errors; so that, if their use has not become general for large numerical calculations, it is because they have not in fact resolved the double problem which the question presents, that of *correctness* in the results, united with *economy* of time.

Struck with similar reflections, Mr Babbage has devoted some years to the realization of a gigantic idea. He proposed to himself nothing less

* This remark seems to require further comment, since it is in some degree calculated to strike the mind as being at variance with the subsequent passage (page 99), where it is explained that *an engine which can effect these four operations* can in fact effect *every species of calculation*. The apparent discrepancy is stronger too in the translation than in the original, owing to its being impossible to render precisely into the English tongue all the niceties of distinction which the French idiom happens to admit of in the phrases used for the two passages we refer to. The explanation lies in this: that in the one case the execution of these four operations is the *fundamental starting-point*, and the object proposed for attainment by the machine is the *subsequent combination of these* in every possible variety; whereas in the other case the execution of some *one* of these four operations, selected at pleasure, is the *ultimatum*, the sole and utmost result that can be proposed for attainment by the machine referred to, and which result it cannot any further combine or work upon. The one *begins* where the other *ends*. Should this distinction not now appear perfectly clear, it will become so on perusing the rest of the memoir, and the notes that are appended to it. Note by translator.

than the construction of a machine capable of executing not merely arithmetical calculations, but even all those of analysis, if their laws are known. The imagination is at first astounded at the idea of such an undertaking; but the more calm reflection we bestow on it, the less impossible does success appear, and it is felt that it may depend on the discovery of some principle so general, that if applied to machinery, the latter may be capable of mechanically translating the operations which may be indicated to it by algebraical notation. The illustrious inventor having been kind enough to communicate to me / some of his views on this subject during a visit he made at Turin, I have, with his approbation, thrown together the impressions they have left on my mind. But the reader must not expect to find a description of Mr Babbage's engine; the comprehension of this would entail studies of much length; and I shall endeavour merely to give an insight into the end proposed, and to develop the principles on which its attainment depends.

I must first premise that this engine is entirely different from that of which there is a notice in the 'Treatise on the Economy of Machinery,' by the same author. But as the latter gave rise* to the idea of the engine in question, I consider it will be a useful preliminary briefly to recall what were Mr Babbage's first essays, and also the circumstances in which they originated.

It is well known that the French Government, wishing to promote the extension of the decimal system, had ordered the construction of logarithmical and trigonometrical tables of enormous extent. M. de Prony, who had been entrusted with the direction of this undertaking, divided it into three sections, to each of which were appointed a special class of persons. In the first section the formulae were so combined as to render them subservient to the purposes of numerical calculation; in the second, these same formulae were calculated for values of the variable, selected at certain successive distances; and under the third section,

* The idea that the one engine is the offspring and has grown out of the other, is an exceedingly natural and plausible supposition, until reflection reminds us that no *necessary* sequence and connection need exist between two such inventions, and that they *may* be wholly independent. M. Menabrea has shared this idea in common with persons who have not his profound and accurate insight into the nature of either engine. In Note A. (see the notes at the end of the memoir) it will be found sufficiently explained, however, that this supposition is unfounded. M. Menabrea's opportunities were by no means such as could be adequate to afford him information on a point like this, which would be naturally and almost unconsciously *assumed*, and would scarcely suggest any enquiry with reference to it. Note by translator.

comprising about eighty individuals, who were most of them only acquainted with the two first rules of arithmetic, the values which were intermediate to those calculated by the second section were interpolated by means of simple additions and subtractions.

An undertaking similar to that just mentioned having been entered upon in England, Mr Babbage conceived that the operations performed under the third section might be executed by a machine; and this idea he realized by means of mechanism, / which has been in part put together, and to which the name Difference Engine is applicable, on account of the principle upon which its construction is founded. To give some notion of this, it will suffice to consider the series of whole square numbers, 1, 4, 9, 16, 25, 36, 49, 64, etc. By subtracting each of these from the succeeding one, we obtain a new series, which we will name the series of first differences, consisting of the numbers 3, 5, 7, 9, 11, 13, 15, etc. On subtracting from each of these the preceding one, we obtain the second differences, which are all constant and equal to 2. We may represent this succession of operations, and their results, in the following table:

	A Column of square numbers	*B* First differences	*C* Second differences	
	1			
		3		
	4		2	*b*
		5		
a	9		2	*d*
		7		
c	16		2	
		9		
	25		2	
		11		
	36			

From the mode in which the two last columns *B* and *C* have been formed, it is easy to see that if, for instance, we desire to pass from the number 5 to the succeeding one 7, we must add to the former the constant difference 2; similarly, if from the square number 9 we would pass to the following one 16, we must add to the former the difference 7, which difference is in other words the preceding difference 5, plus the constant difference 2; or again, which comes to the same thing, to obtain 16 we have only to add together the three numbers 2, 5, 9, placed obliquely in the direction *ab*. Similarly, we obtain the number 25 by summing up the three numbers placed in the oblique direction *dc*: commencing by the addition $2 + 7$, we have the first difference 9 consecutively to 7; adding 16 to the 9 we have the square 25. We see then that the three numbers 2, 5, 9 being given, the whole series of successive square numbers, and that of their first differences likewise, may be obtained by means of simple additions.

Now, to conceive how these operations may be reproduced by a machine, suppose the latter to have three dials, designated as *A*, *B*, *C*, on each of which are traced, say a thousand divisions, by way of example, over which a needle shall pass. The two dials, *C*, *B*, shall have in addition a registering hammer, which is to give a number of strokes equal to that of the divisions indicated by the needle. For each stroke of the registering hammer of the dial *C*, the needle *B* shall advance one division: / similarly, the needle *A* shall advance one division for every stroke of the registering hammer of the dial *B*. Such is the general disposition of the mechanism.

This being understood, let us at the beginning of the series of operations we wish to execute, place the needle *C* on the division 2, the needle *B* on the division 5, and the needle *A* on the division 9. Let us allow the hammer of the dial *C* to strike; it will strike twice, and at the same time the needle *B* will pass over two divisions. The latter will then indicate the number 7, which succeeds the number 5 in the column of first differences. If we now permit the hammer of the dial *B* to strike in its turn, it will strike seven times, during which the needle *A* will advance seven divisions; these added to the nine already marked by it, will give the number 16, which is the square number consecutive to 9. If we now recommence these operations, beginning with the needle *C*, which is always to be left on the division 2, we shall perceive that by repeating them indefinitely, we may successviely reproduce the series of whole square numbers by means of a very simple mechanism.

The theorem on which is based the construction of the machine we have just been describing, is a particular case of the following more general theorem: that if in any polynomial whatever, the highest power of whose variable is *m*, this same variable be increased by equal degrees; the corresponding values of the polynomial then calculated, and the first, second, third, etc., differences of these be taken (as for the preceding series of squares); the *m*th differences will all be equal to each other. So that, in order to reproduce the series of values of the polynomial by means of a machine analogous to the one above described, it is sufficient that there be $(m + 1)$ dials, having the mutual relations we have indicated. As the differences may be either positive or negative, the machine will have a contrivance for either advancing or retrograding each needle, according as the number to be algebraically added may have the sign *plus* or *minus*.

If from a polynomial we pass to a series having an infinite number of terms, arranged according to the ascending powers of the variable, it would at first appear, that in order to apply the machine to the calculation of the function represented by such a series, the mechanism must include an infinite number of dials, which would in fact render the thing impossible. But in many / cases the difficulty will disappear, if we observe that for a great number of functions the series which represent them may be rendered convergent; so that, according to the degree of approximation desired, we may limit ourselves to the calculation of a certain number of terms of the series, neglecting the rest. By this method the question is reduced to the primitive case of a finite polynomial. It is thus that we can calculate the succession of the logarithms of numbers. But since, in this particular instance, the terms which had been originally neglected receive increments in a ratio so continually increasing for equal increments of the variable, that the degree of approximation required would ultimately be affected, it is necessary, at certain intervals, to calculate the value of the function by different methods, and then respectively to use the results thus obtained, as data whence to deduce, by means of the machine, the other intermediate values. We see that the machine here performs the office of the third section of calculators mentioned in describing the tables computed by order of the French Government, and that the end originally proposed is thus fulfilled by it.

Such is the nature of the first machine which Mr Babbage conceived. We see that its use is confined to cases where the numbers required are such as can be obtained by means of simple additions or subtractions; that the machine is, so to speak, merely the expression of one* particular theorem of analysis; and that, in short, its operations cannot be extended so as to embrace the solution of an infinity of other questions included within the domain of mathematical analysis. It was while contemplating the vast field which yet remained to be traversed,

* See Note A

that Mr Babbage, renouncing his original essays, conceived the plan of another system of mechanism whose operations should themselves possess all the generality of algebraical notation, and which, on this account, he denominates the *Analytical Engine.*

Having now explained the state of the question, it is time for me to develop the principle on which is based the construction of this latter machine. When analysis is employed for the solution of any problem, there are usually two classes of operations to execute: firstly, the numerical calculation of the various coefficients; and secondly, their distribution in relation to the quantities affected by them. If, for example, we have to obtain / the product of two binomials $(a + bx)(m + nx)$, the result will be represented by $am + (an + bm)x + bnx^2$, in which expression we must first calculate am, an, bm, bn; then take the sum of $an + bm$; and lastly, respectively distribute the coefficients thus obtained, amongst the powers of the variable. In order to reproduce these operations by means of a machine, the latter must therefore possess two distinct sets of powers: first, that of executing numerical calculations; secondly, that of rightly distributing the values so obtained.

But if human intervention were necessary for directing each of these partial operations, nothing would be gained under the heads of correctness and economy of time; the machine must therefore have the additional requisite of executing by itself all the successive operations required for the solution of a problem proposed to it, when once the *primitive numerical data* for this same problem have been introduced. Therefore, since from the moment that the nature of the calculation to be executed or of the problem to be resolved have been indicated to it, the machine is, by its own intrinsic power, of itself to go through all the intermediate operations which lead to the proposed result, it must exclude all methods of trial and guesswork, and can only admit the direct processes of calculation.*

It is necessarily thus; for the machine is not a thinking being, but simply an automaton which acts according to the laws imposed upon it. This being fundamental, one of the earliest researches its author had to undertake, was that of finding means for effecting the division of one

* This must not be understood in too unqualified a manner. The engine is capable, under certain circumstances, of feeling about to discover which of two or more possible contingencies has occurred, and of then shaping its future course accordingly. Note by translator.

number by another without using the method of guessing indicated by the usual rules of arithmetic. The difficulties of effecting this combination were far from being amongst the least; but upon it depended the success of every other. Under the impossibility of my here explaining the process through which this end is attained, we must limit ourselves to admitting that the four first operations of arithmetic, what is addition, subtraction, multiplication, and division, can be performed in a direct manner through the intervention of the machine. This granted, the machine is thence capable of performing every species of numerical calculation, for all such calculations ultimately resolve themselves into the four operations / we have just named. To conceive how the machine can now go through its functions according to the laws laid down, we will begin by giving an idea of the manner in which it materially represents numbers.

Let us conceive a pile or vertical column consisting of an indefinite number of circular discs, all pierced through their centres by a common axis, around which each of them can take an independent rotatory movement. If round the edge of each of these discs are written the ten figures which constitute our numerical alphabet, we may then, by arranging a series of these figures in the same vertical line, express in this manner any number whatever. It is sufficient for this purpose that the first disc represent units, the second tens, the third hundreds, and so on. When two numbers have been thus written on two distinct columns, we may propose to combine them arithmetically with each other, and to obtain the result on a third column. In general, if we have a series of columns* consisting of discs, which columns we will designate as V_0, V_1, V_2, V_3, V_4, etc., we may require, for instance, to divide the number written on the column V_1 by that on the column V_4, and to obtain the result on the column V_7. To effect this operation, we must impart to the machine two distinct arrangements; through the first it is prepared for executing *a division*, and through the second the columns it is to operate on are indicated to it, and also the column on which the result is to be represented. If this division is to be followed, for example, by the addition of two numbers taken on other columns, the two original arrangements of the machine must be simultaneously altered. If, on the contrary, a series of operations of the same nature is to be gone through, then the first of the original arrangements will remain, and the second alone must be altered. Therefore, the arrangements that may be

* See Note B

communicated to the various parts of the machine, may be distinguished into two principal classes:

First, that relative to the *operations*.

Secondly, that relative to the *Variables*.

By this latter we mean that which indicates the columns to be operated on. As for the operations themselves, they are executed by a special apparatus, which is designated by the name of *mill*, and which itself contains a certain number of columns, similar to those of the Variables. When two numbers are to be / combined together, the machine commences by effacing them from the columns where they are written, that is it places *zero** on every disc of the two vertical lines on which the numbers were represented; and it transfers the numbers to the mill. There, the apparatus having been disposed suitably for the required operation, this latter is effected, and, when completed, the result itself is transferred to the column of Variables which shall have been indicated. Thus the mill is that portion of the machine which works, and the columns of Variables constitute that where the results are represented and arranged. After the preceding explanations, we may perceive that all fractional and irrational results will be represented in decimal fractions. Supposing each column to have forty discs, this extension will be sufficient for all degrees of approximation generally required.

It will now be enquired how the machine can of itself, and without having recourse to the hand of man, assume the successive dispositions suited to the operations. The solution of this problem has been taken from Jacquard's apparatus,† used for the manufacture of brocaded stuffs, in the following manner:

Two species of threads are usually distinguished in woven stuffs; one is the *warp* or longitudinal thread, the other the *woof* or transverse thread, which is conveyed by the instrument called the shuttle, and which crosses the longitudinal thread or warp. When a brocaded stuff is required, it is necessary in turn to prevent certain threads from crossing the woof, and this according to a succession which is determined by the nature of the design that is to be reproduced. Formerly this process was lengthy and difficult, and it was requisite that the workman, by attending to the design which he was to copy, should himself regulate the movements the threads were to take. Thence arose the high price of

* Zero is not *always* substituted when a number is transferred to the mill. This is explained further on in the memoir, and still more fully in Note D. Note by translator.

† See Note C

this description of stuffs, especially if threads of various colours entered into the fabric. To simplify this manufacture, Jacquard devised the plan of connecting each group of threads that were to act together, with a distinct lever belonging exclusively to that group. All these levers terminate in rods, which are united together in one bundle, having usually the form of a parallelopiped with a rectangular base. The rods are cylindrical, and are separated from each other by small intervals. / The process of raising the threads is thus resolved into that of moving these various lever arms in the requisite order. To effect this, a rectangular sheet of pasteboard is taken somewhat larger in size than a section of the bundle of lever arms. If this sheet be applied to the base of the bundle, and an advancing motion be then communicated to the pasteboard, this latter will move with it all the rods of the bundle, and consequently the threads that are connected with each of them. But if the pasteboard, instead of being plain, were pierced with holes corresponding to the extremities of the levers which meet it, then, since each of the levers would pass through the pasteboard during the motion of the latter, they would all remain in their places. We thus see that it is easy so to determine the position of the holes in the pasteboard, that, at any given moment, there shall be a certain number of levers, and consequently of parcels of threads, raised, whilst the rest remain where they were. Supposing this process is successively repeated according to a law indicated by the pattern to be executed, we perceive that this pattern may be reproduced on the stuff. For this purpose we need merely compose a series of cards according to the law required, and arrange them in suitable order one after the other; then, by causing them to pass over a polygonal beam which is so connected as to turn a new face for every stroke of the shuttle, which face shall then be impelled parallelly to itself against the bundle of lever arms, the operation of raising the threads will be regularly performed. Thus we see that brocaded tissues may be manufactured with a precision and rapidity formerly difficult to obtain.

Arrangements analogous to those just described have been introduced into the Analytical Engine. It contains two principal species of cards: first, operation cards, by means of which the parts of the machine are so disposed as to execute any determinate series of operations, such as additions, subtractions, multiplications, and divisions; secondly, cards of the Variables, which indicate to the machine the columns on which the results are to be represented. The cards, when put in motion, successively arrange the various portions of the machine according to the nature of the processes that are to be effected, and the machine at the same time executes these processes by means of the various pieces of mechanism of which it is constituted.

In order more perfectly to conceive the thing, let us select / as an example the resolution of two equations of the first degree with two unknown quantities. Let the following be the two equations, in which x and y are the unknown quantities:

$$\begin{cases} mx + ny = d \\ m'x + n'y = d' \end{cases}$$

We deduce

$$x = \frac{dn' - d'n}{n'm - nm'}$$

and for y an analogous expression. Let us continue to represent by V_0, V_1, V_2, etc., the different columns which contain the numbers, and let us suppose that the first eight columns have been chosen for expressing on them the numbers represented by m, n, d, m', n', d', n, and n', which implies that $V_0 = m$, $V_1 = n$, $V_2 = d$, $V_3 = m'$, $V_4 = n'$, $V_5 = d'$, $V_6 = n$, $V_7 = n'$.

The series of operations commanded by the cards, and the results obtained, may be represented in the following table:

Number of the operations	*Operation cards*	*Cards of the Variables*		*Progress of the operations*
	Symbols indicating the nature of the operations	*Columns on which operations are to be performed*	*Columns which receive results of operations*	
1	$\times$	$V_2 \times V_4 =$	V_8	$= dn'$
2	$\times$	$V_5 \times V_1 =$	V_9	$= d'n$
3	$\times$	$V_4 \times V_0 =$	V_{10}	$= n'm$
4	$\times$	$V_1 \times V_3 =$	V_{11}	$= nm'$
5	$-$	$V_8 - V_9 =$	V_{12}	$= dn' - d'n$
6	$-$	$V_{10} - V_{11} =$	V_{13}	$= n'm - nm'$
7	$\div$	$\frac{V_{12}}{V_{13}} =$	V_{14}	$= x = \frac{dn' - d'n}{n'm - nm'}$

Since the cards do nothing but indicate in what manner and on what columns the machine shall act, it is clear that we must still, in every particular case, introduce the numerical data for the calculation. Thus, in the example we have selected, we must previously inscribe the numerical values of m, n, d, m', n', d', in the order and on the columns indicated, after which the machine when put in action will give the value of the unknown quantity x for this particular case. To obtain the value of y, another series of operations analogous to the preceding must be performed. But we see that they will be only four in number, since the denominator of the expression for y, excepting the sign, is the same as that for x, and equal to $n'm - nm'$. In the preceding table it will be remarked that the column for operations indicates four successive *multiplications*, two *subtractions*, and / one *division*. Therefore, if desired, we need only use three operation cards; to manage which, it is sufficient to introduce into the machine an apparatus which shall, after the first multiplication, for instance, retain the card which relates to this operation, and not allow it to advance so as to be replaced by another one, until after this same operation shall have been four times repeated. In the preceding example we have seen, that to find the value of x we must begin by writing the coefficients m, n, d, m', n', d', upon eight columns, thus repeating n and n' twice. According to the same method, if it were required to calculate y likewise, these coefficients must be written on twelve different columns. But it is possible to simplify this process, and thus to diminish the chances of errors, which chances are greater, the larger the number of the quantities that have to be inscribed previous to setting the machine in action. To understand this simplification, we must remember that every number written on a column must, in order to be arithmetically combined with another number, be effaced from the column on which it is, and transferred to the *mill*. Thus, in the example we have discussed, we will take the two coefficients m and n', which are each of them to enter into *two* different products, that is m into mn' and md', n' into mn' and $n'd$. These coefficients will be inscribed on the columns V_0 and V_4. If we commence the series of operations by the product of m into n', these numbers will be effaced from the columns V_0 and V_4, that they may be transferred to the mill, which will multiply them into each other, and will then command the machine to represent the result, say on the column V_6. But as these numbers are each to be used again in another operation, they must again be inscribed somewhere; therefore, whilst the mill is working out their product, the machine will inscribe them anew on any two columns that may be indicated to it through the cards; and, as in the actual case, there is no reason why they should not resume their former places, we will suppose them again inscribed on V_0 and V_4, whence in short they would not finally disappear, to be reproduced no more, until they should have gone through all the combinations in which they might have to be used.

We see, then, that the whole assemblage of operations requisite for resolving the two* above equations of the first degree, may be definitely represented in the following table: /

Columns on which are inscribed the primitive data	*Number of the operations*	*Cards of the operations*		*Variable cards*			*Statement of results*
		Number of the operation cards	*Nature of each operation*	*Columns acted on by each operation*	*Columns that receive the result of each operation*	*Indication of change of value on any column*	
${}^1V_0 = m$	1	1	×	${}^1V_0 \times {}^1V_4 =$	${}^1V_6 \ldots$	$\begin{Bmatrix} {}^1V_0 = {}^1V_0 \\ {}^1V_4 = {}^1V_4 \end{Bmatrix}$	${}^1V_6 = mn'$
${}^1V_1 = n$	2	”	×	${}^1V_3 \times {}^1V_1 =$	${}^1V_7 \ldots$	$\begin{Bmatrix} {}^1V_3 = {}^1V_3 \\ {}^1V_1 = {}^1V_1 \end{Bmatrix}$	${}^1V_7 = m'n$
${}^1V_2 = d$	3	”	×	${}^1V_2 \times {}^1V_4 =$	${}^1V_8 \ldots$	$\begin{Bmatrix} {}^1V_2 = {}^1V_2 \\ {}^1V_4 = {}^0V_4 \end{Bmatrix}$	${}^1V_8 = dn'$
${}^1V_3 = m'$	4	”	×	${}^1V_5 \times {}^1V_1 =$	${}^1V_9 \ldots$	$\begin{Bmatrix} {}^1V_5 = {}^1V_5 \\ {}^1V_1 = {}^0V_1 \end{Bmatrix}$	${}^1V_9 = d'n$
${}^1V_4 = n'$	5	”	×	${}^1V_0 \times {}^1V_5 =$	${}^1V_{10} \ldots$	$\begin{Bmatrix} {}^1V_0 = {}^0V_0 \\ {}^1V_5 = {}^0V_5 \end{Bmatrix}$	${}^1V_{10} = d'm$
${}^1V_5 = d'$	6	”	×	${}^1V_2 \times {}^1V_3 =$	${}^1V_{11} \ldots$	$\begin{Bmatrix} {}^1V_2 = {}^0V_2 \\ {}^1V_3 = {}^0V_3 \end{Bmatrix}$	${}^1V_{11} = dm'$
	7	2	−	${}^1V_6 - {}^1V_7 =$	${}^1V_{12} \ldots$	$\begin{Bmatrix} {}^1V_6 = {}^0V_6 \\ {}^1V_7 = {}^0V_7 \end{Bmatrix}$	${}^1V_{12} = mn' - m'n$
	8	”	−	${}^1V_8 - {}^1V_9 =$	${}^1V_{13} \ldots$	$\begin{Bmatrix} {}^1V_8 = {}^0V_8 \\ {}^1V_9 = {}^0V_9 \end{Bmatrix}$	${}^1V_{13} = dn' - d'n$
	9	”	−	${}^1V_{10} - {}^1V_{11} =$	${}^1V_{14} \ldots$	$\begin{Bmatrix} {}^1V_{10} = {}^0V_{10} \\ {}^1V_{11} = {}^0V_{11} \end{Bmatrix}$	${}^1V_{14} = d'm - dm'$
	10	3	÷	${}^1V_{13} \div {}^1V_{12} =$	${}^1V_{15} \ldots$	$\begin{Bmatrix} {}^1V_{13} = {}^0V_{13} \\ {}^1V_{12} = {}^1V_{12} \end{Bmatrix}$	${}^1V_{15} = \dfrac{dn' - d'n}{mn' - m'n} = x$
	11	”	÷	${}^1V_{14} \div {}^1V_{12} =$	${}^1V_{16} \ldots$	$\begin{Bmatrix} {}^1V_{14} = {}^0V_{14} \\ {}^1V_{12} = {}^0V_{12} \end{Bmatrix}$	${}^1V_{16} = \dfrac{d'm - dm'}{mn' - m'n} = y$
1	2	3	4	5	6	7	8

* See Note D.

In order to diminish to the utmost the chances of error in inscribing the numerical data of the problem, they are successively placed on one of the columns of the mill; then, by means of cards arranged for this purpose, these same numbers are caused to arrange themselves on the requisite columns, without the operator having to give his attention to it; so that his undivided mind may be applied to the simple inscription of these same numbers.

According to what has now been explained, we see that the collection of columns of Variables may be regarded as a *store* of numbers, accumulated there by the mill, and which, obeying the orders transmitted to the machine by means of the cards, pass alternately from the mill to the store, and from the store to the mill, that they may undergo the transformations demanded by the nature of the calculation to be performed.

Hitherto no mention has been made of the *signs* in the results, and the machine would be far from perfect were it incapable / of expressing and combining amongst each other positive and negative quantities. To accomplish this end, there is, above every column, both of the mill and of the store, a disc, similar to the discs of which the columns themselves consist. According as the digit on this disc is even or uneven, the number inscribed on the corresponding column below it will be considered as positive or negative. This granted, we may, in the following manner, conceive how the signs can be algebraically combined in the machine. When a number is to be transferred from the store to the mill, and *vice versa*, it will always be transferred with its sign, which will be effected by means of the cards, as has been explained in what precedes. Let any two numbers then, on which we are to operate arithmetically, be placed in the mill with their respective signs. Suppose that we are first to add them together; the operation cards will command the addition: if the two numbers be of the same sign, one of the two will be entirely effaced from where it was inscribed, and will go to add itself on the column which contains the other number; the machine will, during this operation, be able, by means of a certain apparatus, to prevent any movement in the disc of signs which belongs to the column on which the addition is made, and thus the result will remain with the sign which the two given numbers originally had. When two numbers have two different signs, the addition commanded by the card will be changed into a subtraction through the intervention of mechanisms which are brought into play by this very difference of sign. Since the subtraction can only be effected on the larger of the two numbers, it must be arranged that the disc of signs of the larger number shall not move whilst the smaller of the two numbers is being effaced from its column and subtracted from the other, whence the result will have the sign of this latter, just as in fact it ought to be. The combinations to which algebraical subtraction give rise, are analogous to the preceding. Let us pass on to multiplication. When two numbers to be multiplied are of the same sign, the result is positive; if the signs are different, the product must be negative. In order that the machine may act conformably to this law, we have but to conceive that on the column containing the product of the two given numbers, the digit which indicates the sign of that product, has been formed by the mutual addition of the two digits that respectively indicated the signs of the two given numbers; it is then obvious / that if the digits of the signs are both even, or both odd, their sum will be an even number, and consequently will express a positive number; but that if, on the contrary, the two digits of the signs are one even and the other odd, their sum will be an odd number, and will consequently express a negative number. In the case of division, instead of adding the digits of the discs, they must be subtracted one from the other, which will produce results analogous to the preceding; that is to say, that if these figures are both even or both uneven, the remainder of this subtraction will be even; and it will be uneven in the contrary case. When I speak of mutually adding or subtracting the numbers expressed by the digits of the signs, I merely mean that one of the sign discs is made to advance or retrograde a number of divisions equal to that which is expressed by the digit on the other sign disc. We see, then, from the preceding explanation, that it is possible mechanically to combine the signs of quantities so as to obtain results conformable to those indicated by algebra.*

The machine is not only capable of executing those numerical calculations which depend on a given algebraical formula, but it is also fitted for analytical calculations in which there are one or several variables to be considered. It must be assumed that the analytical expression to be operated on can be developed according to powers of the variable, or according to determinate functions of this same variable, such as circular functions, for instance; and similarly for the

* Not having had leisure to discuss with Mr Babbage the manner of introducing into his machine the combination of algebraical signs, I do not pretend here to expose the method he uses for this purpose; but I considered that I ought myself to supply the deficiency, conceiving that this paper would have been imperfect if I had omitted to point out one means that might be employed for resolving this essential part of the problem in question.

result that is to be attained. If we then suppose that above the columns of the store, we have inscribed the powers or the functions of the variable, arranged according to whatever is the prescribed law of development, the coefficients of these several terms may be respectively placed on the corresponding column below each. In this manner we shall have a representation of an analytical development; and, supposing the position of the several terms composing it to be invariable, the problem will be reduced to that of calculating their coefficients according to the laws demanded by the nature of the question. In order to make this more clear, we shall take the / following* very simple example, in which we are to multiply $(a + bx^1)$ by $(A + B\cos^1 x)$. We shall begin by writing x^0, x^1, $\cos^0 x$, $\cos^1 x$, above the columns V_0, V_1, V_2, V_3; then, since from the form of the two functions to be combined, the terms which are to compose the products will be of the following nature, $x^0 \cdot \cos^0 x$, $x^0 \cdot \cos^1 x$, $x^1 \cdot \cos^0 x$, $x^1 \cdot \cos^1 x$; these will be inscribed about the columns V_4, V_5, V_6, V_7. The coefficients of x^0, x^1, $\cos^0 x$, $\cos^1 x$ being given, they will, by means of the mill, be passed to the columns V_0, V_1, V_2 and V_3. Such are the primitive data of the problem. It is now the business of the machine to work out its solution, that is to find the coefficients which are to be inscribed on V_4, V_5, V_6, V_7. To attain this object, the law of formation of these same coefficients being known, the machine will act through the intervention of the cards, in the manner indicated by the table on page 108.

It will now be perceived that a general application may be made of the principle developed in the preceding example, to every species of process which it may be proposed to effect on series submitted to calculation. It is sufficient that the law of formation of the coefficients be known, and that this law be inscribed on the cards of the machine, which will then of itself execute all / the calculations requisite for arriving at the proposed result. If, for instance, a recurring series were proposed, the law of formation of the coefficients being here uniform, the same operations which must be performed for one of them will be repeated for all the others; there will merely be a change in the locality of the operation, that is it will be performed with different columns. Generally, since every analytical expression is susceptible of being expressed in a series ordered according to certain functions of the variable, we perceive that the machine will include all analytical calculations which can be definitively reduced to the formation of

* See Note E.

* Columns above which are written the functions of the variable		Coefficients: Given	Coefficients: To be formed	Cards of the operations: Number of the operations	Cards of the operations: Nature of the operation	Cards of the variables: Columns on which operations are to be performed	Cards of the variables: Columns on which are to be inscribed the results of the operations	Cards of the variables: Indication of change of value on any column submitted to an operation	Cards of the variables: Results of the operations
x^0	$^1V_0$	a							
x^1	$^1V_1$	b							
$\cos^0 x$	$^1V_2$	A							
$\cos^1 x$	$^1V_3$	B							
$x^0 \cos^0 x$	$^0V_4$		aA	1	$\times$	$^1V_0 \times {}^1V_2 =$	$^1V_4$	$\left\{\begin{matrix} ^1V_0 = {}^1V_0 \\ ^1V_2 = {}^1V_2 \end{matrix}\right\}$	$^1V_4 = aA$ coefficients of $x^0 \cos^0 x$
$x^0 \cos^1 x$	$^0V_5$		aB	2	$\times$	$^1V_0 \times {}^1V_3 =$	$^1V_5 \ldots$	$\left\{\begin{matrix} ^1V_0 = {}^0V_0 \\ ^1V_3 = {}^1V_3 \end{matrix}\right\}$	$^1V_5 = aB$ " " $x^0 \cos^1 x$
$x^1 \cos^0 x$	$^0V_6$		bA	3	$\times$	$^1V_1 \times {}^1V_2 =$	$^1V_6$	$\left\{\begin{matrix} ^1V_1 = {}^1V_1 \\ ^1V_2 = {}^0V_2 \end{matrix}\right\}$	$^1V_6 = bA$ " " $x^1 \cos^0 x$
$x^1 \cos^1 x$	$^0V_7$		bB	4	$\times$	$^1V_1 \times {}^1V_3 =$	$^1V_7$	$\left\{\begin{matrix} ^1V_1 = {}^0V_1 \\ ^1V_3 = {}^0V_3 \end{matrix}\right\}$	$^1V_7 = bB$ " " $x^1 \cos^1 x$

coefficients according to certain laws, and to the distribution of these with respect to the variables.

We may deduce the following important consequence from these explanations, viz. that since the cards only indicate the nature of the operations to be performed, and the columns of Variables with which they are to be executed, these cards will themselves possess all the generality of analysis, of which they are in fact merely a translation. We shall now further examine some of the difficulties which the machine must surmount, if its assimilation to analysis is to be complete. There

* For an explanation of the upper left-hand indices attached to the *V*'s in this and in the preceding Table, we must refer the reader to Note D, amongst those appended to the memoir.—Note by translator.

are certain functions which necessarily change in nature when they pass through zero or infinity, or whose values cannot be admitted when they pass these limits. When such cases present themselves, the machine is able, by means of a bell, to give notice that the passage through zero or infinity is taking place, and it then stops until the attendant has again set it in action for whatever process it may next be desired that it shall perform. If this process has been foreseen, then the machine, instead of ringing, will so dispose itself as to present the new cards which have relation to the operation that is to succeed the passage through zero and infinity. These new cards may follow the first, but may only come into play contingently upon one or other of the two circumstances just mentioned taking place.

Let us consider a term of the form ab^n; since the cards are but a translation of the analytical formula, their number in this particular case must be the same, whatever be the value of n; that is to say, whatever be the number of multiplications required for elevating b to the nth power (we are supposing for the moment that n is a whole number). Now, since the exponent n indicates that b is to be multiplied n times by itself, and all these operations are of the same nature, it will be sufficient to employ / one single operation card, viz. that which orders the multiplication.

But when n is given for the particular case to be calculated, it will be further requisite that the machine limit the number of its multiplications according to the given values. The process may be thus arranged. The three numbers a, b, and n will be written on as many distinct columns of the store; we shall designate them V_0, V_1, V_2; the result ab^n will place itself on the column V_3. When the number n has been introduced into the machine, a card will order a certain registering-apparatus to mark $(n - 1)$, and will at the same time execute the multiplication of b by b. When this is completed, it will be found that the registering apparatus has effaced a unit, and that it only marks $(n - 2)$; while the machine will now again order the number b written on the column V_1 to multiply itself with the product b^2 written on the column V_3, which will give b^3. Another unit is then effaced from the registering apparatus, and the same processes are continually repeated until it only marks zero. Thus the number b^n will be found inscribed on V_3, when the machine, pursuing its course of operations, will order the product of b^n by a; and the required calculation will have been completed without there being any necessity that the number of operation cards used should vary with the value of n. If n were negative, the cards, instead of ordering the multiplication of a by b^n, would order its division; this we can easily conceive, since every number, being inscribed with its respective sign, is consequently capable of reacting on the nature of the operations to be executed. Finally, if n were fractional, of the form p/q, an additional column would be used for the inscription of q, and the machine would bring into action two sets of processes, one for raising b to the power p, the other for extracting the qth root of the number so obtained.

Again, it may be required, for example, to multiply an expression of the form $ax^m + bx^n$ by another $Ax^p + Bx^q$, and then to reduce the product to the least number of terms, if any of the indices are equal. The two factors being ordered with respect to x, the general result of the multiplication would be

$$Aax^{m+p} + Abx^{n+p} + Bax^{m+q} + Bbx^{n+q}.$$

Up to this point the process presents no difficulties; but suppose that we have $m = p$ and $n = q$, and that we wish to reduce the two middle terms to / a single one $(Ab + Ba)x^{m+q}$. For this purpose, the cards may order $m + q$ and $n + p$ to be transferred into the mill, and there subtracted one from the other; if the remainder is nothing, as would be the case on the present hypothesis, the mill will order other cards to bring to it the coefficients Ab and Ba, that it may add them together and give them in this state as a coefficient for the single term $x^{n+p} = x^{m+q}$.

This example illustrates how the cards are able to reproduce all the operations which intellect performs in order to attain a determinate result, if these operations are themselves capable of being precisely defined.

Let us now examine the following expression:

$$2\cdot\frac{2^2\cdot 4^2\cdot 6^2\cdot 8^2\cdot 10^2\cdot\ldots\cdot(2n)^2}{1^2\cdot 3^2\cdot 5^2\cdot 7^2\cdot 9^2\cdot\ldots\cdot(2n-1)^2\cdot(2n+1)^2}$$

which we know becomes equal to the ratio of the circumference to the diameter, when n is infinite. We may require the machine not only to perform the calculation of this fractional expression, but further to give indication as soon as the value becomes identical with that of the ratio of the circumference to the diameter when n is infinite, a case in which the computation would be impossible. Observe that we should thus require of the machine to interpret a result not of itself evident, and that this is not amongst its attributes, since it is no thinking being.

Nevertheless, when the cos of $n = \infty$ has been foreseen,[a] a card may immediately order the substitution of the value of π (π being the ratio of the circumference to the diameter), without going through the series of calculations indicated. This would merely require that the machine contain a special card, whose office it should be to place the number π in a direct and independent manner on the column indicated to it. And here we should introduce the mention of a third species of cards, which may be called *cards of numbers*. There are certain numbers, such as those expressing the ratio of the circumference to the diameter, the Numbers of Bernoulli, etc., which frequently present themselves in calculations. To avoid the necessity for computing them every time they have to be used, certain cards may be combined specially in order to give these numbers ready made into the mill, whence they afterwards go and place themselves on those columns of the store that are destined for them. Through this means the machine will be susceptible of those simplifications / afforded by the use of numerical tables. It would be equally possible to introduce, by means of these cards, the logarithms of numbers; but perhaps it might not be in this case either the shortest or the most appropriate method; for the machine might be able to perform the same calculations by other more expeditious combinations, founded on the rapidity with which it executes the four first operations of arithmetic. To give an idea of this rapidity, we need only mention that Mr Babbage believes he can, by his engine, form the product of two numbers, each containing twenty figures, in *three minutes*.

Perhaps the immense number of cards required for the solution of any rather complicated problem may appear to be an obstacle; but this does not seem to be the case. There is no limit to the number of cards that can be used. Certain stuffs require for their fabrication not less than *twenty thousand* cards, and we may unquestionably far exceed even this quantity.*

Resuming what we have explained concerning the Analytical Engine,

* See Note F.

---

[a] *Sic*; this is a literal translation of a misprint in Menabrea's original paper (see p. 79 in this volume). The phrase 'the cos of $n = \infty$' should read 'the case of $n = \infty$'. Some recent commentators have cited this howler as evidence of the modest level of Ada Lovelace's mathematical attainment. It is interesting to note that the error lay uncorrected in two reprintings (H. P. Babbage's *Babbage's calculating engines* (London, 1889) and P. and E. Morrison's *Charles Babbage and his calculating engines* (New York, 1961)). The error was silently corrected in B. V. Bowden's careful reprinting in *Faster than thought* (London, 1953). (*Reference*: Dorothy Stein, *Ada: a life and its legacy* (Cambridge, Ma., 1985).)

we may conclude that it is based on two principles: the first, consisting in the fact that every arithmetical cauculation ultimately depends on four principal operations – addition, subtraction, multiplication, and division; the second, in the possibility of reducing every analytical calculation to that of the coefficients for the several terms of a series. If this last principle be true, all the operations of analysis come within the domain of the engine. To take another point of view: the use of the cards offers a generality equal to that of algebraical formulae, since such a formula simply indicates the nature and order of the operations requisite for arriving at a certain definite result, and similarly the cards merely command the engine to perform these same operations; but in order that the mechanisms may be able to act to any purpose, the numerical data of the problem must in every particular case be introduced. Thus the same series of cards will serve for all questions whose sameness of nature is such as to require nothing altered excepting the numerical data. In this light the cards are merely a translation of algebraical formulae, or, to express it better, another form of analytical notation.

Since the engine has a mode of acting peculiar to itself, it will in every particular case be necessary to arrange the series of calculations conformably to the means which the machine possesses; / for such or such a process which might be very easy for a calculator, may be long and complicated for the engine, and *vice versa.*

Considered under the most general point of view, the essential object of the machine being to calculate, according to the laws dictated to it, the values of numerical coefficients which it is then to distribute appropriately on the columns which represent the variables, it follows that the interpretion of formulae and of results is beyond its province, unless indeed this very interpretation be itself susceptible of expression by means of the symbols which the machine employs. Thus, although it is not itself the being that reflects, it may yet be considered as the being which executes the conceptions of intelligence.* The cards receive the impress of these conceptions, and transmit to the various trains of mechanism composing the engine the orders necessary for their action. When once the engine shall have been constructed, the difficulty will be reduced to the making out of the cards; but as these are merely the translation of algebraical formulae, it will, by means of some simple notations, be easy to consign the execution of them to a workman. Thus

* See Note G.

the whole intellectual labour will be limited to the preparation of the formulae, which must be adapted for calculation by the engine.

Now, admitting that such an engine can be constructed, it may be enquired: what will be its utility? To recapitulate; it will afford the following advantages: First, rigid accuracy. We know that numerical calculations are generally the stumbling-block to the solution of problems, since errors easily creep into them, and it is by no means always easy to detect these errors. Now the engine, by the very nature of its mode of acting, which requires no human intervention during the course of its operations, presents every species of security under the head of correctness; besides, it carries with it its own check; for at the end of every operation it prints off, not only the results, but likewise the numerical data of the question; so that it is easy to verify whether the question has been correctly proposed. Secondly, economy of time: to convince ourselves of this, we need only recollect that the multiplication of two numbers, consisting each of twenty figures, requires at the very utmost three minutes. Likewise, when a long series of identical computations is to be performed, such as those required for the formation of / numerical tables, the machine can be brought into play so as to give several results at the same time, which will greatly abridge the whole amount of the processes. Thirdly, economy of intelligence: a simple arithmetical computation requires to be performed by a person possessing some capacity; and when we pass to more complicated calculations, and wish to use algebraical formulae in particular cases, knowledge must be possessed which presupposes preliminary mathematical studies of some extent. Now the engine, from its capability of performing by itself all these purely material operations, spares intellectual labour, which may be more profitably employed. Thus the engine may be considered as a real manufactory of figures, which will lend its aid to those many useful sciences and arts that depend on numbers. Again, who can foresee the consequences of such an invention? In truth, how many precious observations remain practically barren for the progress of the sciences, because there are not powers sufficient for computing the results! And what discouragement does the perspective of a long and arid computation cast into the mind of a man of genius, who demands time exclusively for meditation, and who beholds it snatched from him by the material routine of operations! Yet it is by the laborious route of analysis that he must reach truth; but he cannot pursue this unless guided by numbers; for without numbers it is not given us to raise the veil which envelopes the mysteries of nature.

Thus the idea of constructing an apparatus capable of aiding human weakness in such researches, is a conception which, being realized, would mark a glorious epoch in the history of the sciences. The plans have been arranged for all the various parts, and for all the wheelwork, which compose this immense apparatus, and their action studied; but these have not yet been fully combined together in the drawings* and mechanical notation.† The confidence which the genius of Mr Babbage must inspire, affords legitimate ground for hope that this enterprise will be crowned with success; and whilst we render homage to the intelligence which directs it, let us breathe aspirations for the accomplishment of such an undertaking. /

NOTES BY THE TRANSLATOR

Note A. Page 97

The particular function whose integral the Difference Engine was constructed to tabulate, is

$$\Delta^7 u_z = 0.$$

The purpose which that engine has been specially intended and adapted to fulfil, is the computation of nautical and astronomical tables. The integral of

$$\Delta^7 u_z = 0$$

being

$$u_z = a + bx + cx^2 + dx^3 + ex^4 + fx^5 + gx^6$$

the constants *a*, *b*, *c*, etc., are represented on the seven columns of discs, of which the engine consists. It can therefore tabulate *accurately* and to an *unlimited extent*, all series whose general term is comprised in the above formula; and it can also tabulate *approximatively* between

* This sentence has been slightly altered in the translation in order to express more exactly the present state of the engine. Note by translator.

† The notation here alluded to is a most interesting and important subject, and would have well deserved a separate and detailed note upon it, amongst those appended to the memoir. It has, however, been impossbile, within the space allotted, even to touch upon so wide a field. Note by translator.

intervals of greater or less extent, all other series which are capable of tabulation by the method of differences.

The Analytical Engine, on the contrary, is not merely adapted for *tabulating* the results of one particular function and of no other, but for *developing and tabulating* any function whatever. In fact the engine may be described as being the material expression of any indefinite function of any degree of generality and complexity, such as for instance,

$$F(x, y, z, \log x, \sin y, x^p, \text{etc.})$$

which is, it will be observed, a function of all other possible functions of any number of quantities.

In this, which we may call the *neutral* or *zero* state of the engine, it is ready to receive at any moment, by means of cards constituting a portion of its mechanism (and applied on the principle of those used in the Jacquard loom), the impress of whatever *special* function we may desire to develop or to tabulate. These cards contain within themselves (in a manner explained in the memoir itself, pages 100 and 101) the law of development of the particular function that may be under consideration, and they compel the mechanism to act accordingly in a certain corresponding order. One of the simplest cases would be, for example, to suppose that

$$F(x, y, z, \text{etc., etc.})$$

is the particular function

$$\Delta^n u_z = 0$$

which the Difference Engine tabulates for values of n only up to 7. In this case the cards would order the mechanism to go through that succession of operations which would tabulate

$$u_z = a + bx + cx^2 + \ldots + mx^{n-1}$$

where n might be any number whatever. /

These cards, however, have nothing to do with the regulation of the particular *numerical* data. They merely determine the *operations** to be

* We do not mean to imply that the *only* use made of the Jacquard cards is that of regulating the algebraical *operations*. But we mean to explain that *those* cards and portions of mechanism which regulate those *operations*, are wholly independent of those which are used for other purposes. M. Menabrea explains that there are *three* classes of cards used in the engine for three distinct sets of objects, viz. *Cards of the Operations*, *Cards of the Variables*, and certain *Cards of Numbers*. (See pages 101 and 111.)

effected, which operations may of course be performed on an infinite variety of particular numerical values, and do not bring out any definite numerical results unless the numerical data of the problem have been impressed on the requisite portions of the train of mechanism. In the above example, the first essential step towards an arithmetical result, would be the substitution of specific numbers for n, and for the other primitive quantities which enter into the function.

Again, let us suppose that for F we put two complete equations of the fourth degree between x and y. We must then express on the cards the law of elimination for such equations. The engine would follow out those laws, and would ultimately give the equation of one variable which results from such elimination. Various *modes* of elimination might be selected; and of course the cards must be made out accordingly. The following is one mode that might be adopted. The engine is able to multiply together any two functions of the form

$$a + bx + cx^2 + \ldots + px^n.$$

This granted, the two equations may be arranged according to the powers of y, and the coefficients of the powers of y may be arranged according to powers of x. The elimination of y will result from the successive multiplications and subtractions of several such functions. In this, and in all other instances, as was explained above, the particular *numerical* data and the *numerical* results are determined by means and by portions of the mechanism which act quite independently of those that regulate the *operations.*

In studying the action of the Analytical Engine, we find that the peculiar and independent nature of the considerations which in all mathematical analysis belong to *operations*, as distinguished from *the objects operated upon* and from the *results* of the operations performed upon those objects, is very strikingly defined and separated.

It is well to draw attention to this point, not only because its full appreciation is essential to the attainment of any very just and adequate general comprehension of the powers and mode of action of the Analytical Engine, but also because it is one which is perhaps too little kept in view in the study of mathematical science in general. It is, however, impossible to confound it with other considerations, either when we trace the manner in which that engine attains its results, or when we prepare the data for its attainment of those results. It were much to be desired, that when mathematical processes pass through the human brain instead of through the medium of inanimate mechanism, it were equally a necessity of things that the reasonings connected with *operations* should hold the same just place as a clear and well-defined branch of the subject of analysis, a fundamental but yet independent / ingredient in the science, which they must do in studying the engine. The confusion, the difficulties, the contradictions which, in consequence of a want of accurate distinctions in this particular, have up to even a recent period encumbered mathematics in all those branches involving the consideration of negative and impossible quantities, will at once occur to the reader who is at all versed in this science, and would alone suffice to justify dwelling somewhat on the point, in connection with any subject so peculiarly fitted to give forcible illustration of it, as the Analytical Engine. It may be desirable to explain, that by the word *operation*, we mean *any process which alters the mutual relation of two or more things*, be this relation of what kind it may. This is the most general definition, and would include all subjects in the universe. In abstract mathematics, of course operations alter those particular relations which are involved in the considerations of number and space, and the *results* of operations are those peculiar results which correspond to the nature of the subjects of operation. But the science of operations, as derived from mathematics more especially, is a science of itself, and has its own abstract truth and value; just as logic has its own peculiar truth and value, independently of the subjects to which we may apply its reasonings and processes. Those who are accustomed to some of the more modern views of the above subject, will know that a few fundamental relations being true, certain other combinations of relations must of necessity follow; combinations unlimited in variety and extent if the deductions from the primary relations be carried on far enough. They will also be aware that one main reason why the separate nature of the science of operations has been little felt, and in general little dwelt on, is the *shifting* meaning of many of the symbols used in mathematical notation. First, the symbols of *operation* are frequently *also* the symbols of the *results* of operations. We may say that these symbols are apt to have both a *retrospective* and a *prospective* signification. They may signify either relations that are the consequence of a series of processes already performed, or relations that are yet to be effected through certain processes. Secondly, figures, the symbols of *numerical magnitude*, are frequently *also* the symbols of *operations*, as when they are the indices of powers. Wherever terms have a shifting meaning, independent sets of considerations are liable to become complicated together, and reasonings and results are frequently falsified. Now in the Analytical Engine the operations which come under the first of the above heads, are ordered and combined by means of a notation and of a train of mechanism which belong exclusively to themselves; and with respect to the second head, whenever numbers meaning *operations* and not *quantities* (such as the indices of powers), are inscribed on any column or set of columns, those columns immediately act in a wholly separate and independent manner, becoming connected with the *operating mechanism* exclusively, and re-acting upon this. They never come into combination with numbers upon any other columns meaning *quantities*; though, of course, if there are numbers meaning *operations* upon *n* columns, these may *combine amongst each other*, and will often be required to do so, just as numbers meaning *quantities* combine with each other in any variety. It might have been arranged that all numbers meaning *operations* should have appeared on some separate portion of the engine from that which presents / numerical *quantities*; but the present mode is in some cases more simple, and offers in reality quite as much distinctness when understood.

The operating mechanism can even be thrown into action independently of any object to operate upon (although of course no *result* could then be developed). Again, it might act upon other things besides *number*, were objects found whose mutual fundamental relations could be expressed by those of the abstract science of operations, and which should be also susceptible of adaptations to the action of the operating notation and mechanism of the engine. Supposing, for instance, that the fundamental relations of pitched sounds in the science of harmony and of musical composition were susceptible of such expression and adaptations, the engine might compose elaborate and scientific pieces of music of any degree of complexity or extent.

The Analytical Engine is an *embodying of the science of operations*, constructed with peculiar reference to abstract number as the subject of those operations. The Difference Engine is the embodying of *one particular and very limited set of operations*, which (see the notation used in note B) may be expressed thus, (+, +, +, +, +, +), or thus, 6 (+). Six repetitions of the one operation, +, is, in fact, the whole sum and object of that engine. It has seven columns, and a number on any column can add itself to a number on the next column to its *right hand*. So that, begining with the column furthest to the left, six additions can be effected, and the result appears on the seventh column, which is the last on the right hand. The *operating* mechanism of this engine acts in as separate and independent a manner as that of the Analytical Engine; but being susceptible of only one unvarying and restricted combination, it has little force or interest in illustration of the distinct nature of the *science of operations*. The importance of regarding the Analytical Engine under this point of view will, we think, become more and more obvious, as the reader proceeds with M. Menabrea's clear and masterly article. The calculus of operations is likewise in itself a topic of so much interest, and has of late years been so much more written on and thought on than formerly, that any bearing which that engine, from its mode of constitution, may possess upon the illustration of this branch of mathematical science, should not be overlooked. Whether the inventor of this engine had any such views in his mind whilst working out the invention, or whether he may subsequently ever have regarded it under this phase, we do not know; but it is one that forcibly occurred to ourselves on becoming acquainted with the means through which analytical combinations are actually attained by the mechanism. We cannot forbear suggesting one practical result which it appears to us must be greatly facilitated by the independent manner in which the engine orders and combines its *operations*: we allude to the attainment of those combinations into which *imaginary quantities* enter. This is a branch of its processes into which we have not had the opportunity of enquiring, and our conjecture therefore as to the principle on which we conceive the accomplishment of such results may have been made to depend, is very probably not in accordance with the fact, and less subservient for the purpose than some other principles, or at least requiring the cooperation of others. It seems to us obvious, however, that where operations are so independent in their mode of acting, it must be easy by means of a few simple / provisions and additions in arranging the mechanism, to bring out a *double* set of *results*, viz. first, the *numerical magnitudes* which are the results of operations performed on *numerical data*. (These results are the *primary* object of the engine). Secondly, the *symbolical results* to be attached to those numerical results, which symbolical results are not less the necessary and logical consequences of operations performed upon *symbolical data*, than are numerical results when the data are numerical.*

If we compare together the powers and the principles of construction

* In fact such an extension as we allude to, would merely constitute a further and more perfected development of any system introduced for making the proper combinations of the signs *plus* and *minus*. How ably M. Menabrea has touched on this restricted case is pointed out in Note B.

of the Difference and of the Analytical Engines, we shall perceive that the capabilities of the latter are immeasurably more extensive than those of the former, and that they in fact hold to each other the same relationship as that of analysis to arithmetic. The Difference Engine can effect but one particular series of operations, viz. that required for tabulating the integral of the special function

$$\Delta^n u_z = 0$$

and as it can only do this for values of n up to 7,* it cannot be considered as being the most *general* expression even of *one particular* function, much less as being the expression of any and all possible functions of all degrees of generality. The Difference Engine can in reality (as has been already partly explained) do nothing but *add*; and any other processes, not excepting those of simple subtraction, multiplication, and division, can be performed by it only just to that extent in which it is possible, by judicious mathematical arrangement and artifices, to reduce them to a *series of additions*. The method of differences is, in fact, a method of additions; and as it includes within its means a larger number of results attainable by *addition* simply, than any other mathematical principle, it was very appropriately selected as the basis on which to construct *an adding machine*, so as to give to the powers of such a machine the widest possible range. The Analytical Engine, on the contrary, can either add, subtract, multiply, or divide with equal facility; and performs each of these four operations in a direct manner, without the aid of any of the other three. This one fact implies everything; and it is scarcely necessary to point out, for instance, that while the Difference Engine can merely *tabulate*, / and is incapable of *developing*, the Analytical Engine can *either tabulate or develop*.

The former engine is in its nature strictly *arithmetical*, and the results it can arrive at lie within a very clearly defined and restricted range,

* The machine might have been constructed so as to tabulate for a higher value of n than seven. Since, however, every unit added to the value of n increases the extent of the mechanism requisite, there would on this account be a limit beyond which it could not be practically carried. Seven is sufficiently high for the calculation of all ordinary tables.

The fact that, in the Analytical Engine, the same extent of mechanism suffices for the solution of $\Delta^n u_z = 0$, whether $n = 7$, $n = 100{,}000$, or $n =$ any number whatever, at once suggests how entirely distinct must be the *nature of the principles* through whose application matter has been enabled to become the working agent of abstract mental operations in each of these engines respectively; and it affords an equally obvious presumption, that in the case of the Analytical Engine, not only are those principles in themselves of a higher and more comprehensive description, but also such as must vastly extend the *practical* value of the engine whose basis they constitute.

while there is no finite line of demarcation which limits the powers of the Analytical Engine. These powers are co-extensive with our knowledge of the laws of analysis itself, and need be bounded only by our acquaintance with the latter. Indeed we may consider the engine as the *material and mechanical representative* of analysis, and that our actual working powers in this department of human study will be enabled more effectually than heretofore to keep pace with our theoretical knowledge of its principles and laws, through the complete control which the engine gives us over the *executive manipulation* of algebraical and numerical symbols.

Those who view mathematical science not merely as a vast body of abstract and immutable truths, whose intrinsic beauty, symmetry, and logical completeness, when regarded in their connection together as a whole, entitle them to a prominent place in the interest of all profound and logical minds, but as possessing a yet deeper interest for the human race, when it is remembered that this science constitutes the language through which alone we can adequately express the great facts of the natural world, and those unceasing changes of mutual relationship which, visibly or invisibly, consciously or unconsciously to our immediate physical perceptions, are interminably going on in the agencies of the creation we live amidst: those who thus think on mathematical truth as the instrument through which the weak mind of man can most effectively read his Creator's works, will regard with especial interest all that can tend to facilitate the translation of its principles into explicit practical forms.

The distinctive characteristic of the Analytical Engine, and that which has rendered it possible to endow mechanism with such extensive faculties as bid fair to make this engine the executive right hand of abstract algebra, is the introduction into it of the principle which Jacquard devised for regulating, by means of punched cards, the most complicated patterns in the fabrication of brocaded stuffs. It is in this that the distinction between the two engines lies. Nothing of the sort exists in the Difference Engine. We may say most aptly that the Analytical Engine *weaves algebraical patterns* just as the Jacquard loom weaves flowers and leaves. Here, it seems to us, resides much more of originality than the Difference Engine can be fairly entitled to claim. We do not wish to deny to this latter all such claims. We believe that it is the only proposal or attempt ever made to construct a calculating machine *founded on the principle of successive orders of differences*, and capable of *printing off its own results*; and that this engine surpasses its predecessors, both in the extent of the calculations which it can perform, in the facility, certainty, and accuracy with which it can effect them, and in the absence of all necessity for the intervention of human intelligence *during the performance of its calculations.* Its nature is, however, limited to the strictly arithmetical, and it is far from being the first or only scheme for constructing *arithmetical* calculating machines with more or less of success.

The bounds of *arithmetic* were however outstepped the moment the / idea of applying the cards had occurred; and the Analytical Engine does not occupy common ground with mere 'calculating machines'. It holds a position wholly its own; and the considerations it suggests are most interesting in their nature. In enabling mechanism to combine together *general* symbols, in successions of unlimited variety and extent, a uniting link is established between the operations of matter and the abstract mental processes of the *most abstract* branch of mathematical science. A new, a vast, and a powerful language is developed for the future use of analysis, in which to wield its truths so that these may become of more speedy and accurate practical application for the purposes of mankind than the means hitherto in our possession have rendered possible. Thus not only the mental and the material, but the theoretical and the practical in the mathematical world, are brought into more intimate and effective connection with each other. We are not aware of its being on record that anything partaking in the nature of what is so well designated the *Analytical* Engine has been hitherto proposed, or even thought of, as a practical possibility, any more than the idea of a thinking or of a reasoning machine.

We will touch on another point which constitutes an important distinction in the modes of operating of the Difference and Analytical Engines. In order to enable the former to do its business, it is necessary to put into its columns the series of numbers constituting the first terms of the several orders of differences for whatever is the particular table under consideration. The machine then works *upon* these as its data. But these data must themsevles have been already computed through a series of calculations by a human head. Therefore that engine can only produce results depending on data which have been arrived at by the explicit and actual working out of processes that are in their nature different from any that come within the sphere of its own powers. In other words, an *analysing* process must have been gone through by a human mind in order to obtain the data upon which the engine then *synthetically* builds its results. The Difference Engine is in its character exclusively *synthetical*, while the Analytical Engine is equally capable of analysis or of synthesis.

It is true that the Difference Engine can calculate to a much greater extent with these few preliminary data, than the data themselves required for their own determination. The table of squares, for instance, can be calculated to any extent whatever, when the numbers *one* and *two* are furnished; and a very few differences computed at any part of a table of logarithms would enable the engine to calculate many hundreds or even thousands of logarithms. Still the circumstance of its requiring, as a previous condition, that any function whatever shall have been numerically worked out, makes it very inferior in its nature and advantages to an engine which, like the Analytical Engine, requires merely that we should know the *succession and distribution of the operations* to be performed; without there being any occasion,* in order to obtain data on which it can work, for our ever having gone through either the same particular operations which it is itself to effect, or any others. Numerical data must of course be given it, but they are mere arbitrary ones; not data that could only be arrived at through a systematic and necessary series of previous numerical calculations, which is quite a different thing. /

To this it may be replied that an analysing process must equally have been performed in order to furnish the Analytical Engine with the necessary *operative* data; and that herein may also lie a possible source of error. Granted that the actual mechanism is unerring in its processes, the *cards* may give it wrong orders. This is unquestionably the case; but there is much less chance of error, and likewise far less expenditure of time and labour, where operations only, and the distribution of these operations, have to be made out, than where explicit numerical results are to be attained. In the case of the Analytical Engine we have undoubtedly to lay out a certain capital of analytical labour in one particular line; but this is in order that the engine may bring us in a much larger return in another line. It should be remembered also that the cards when once made out for any formula, have all the generality of algebra, and include an infinite number of particular cases.

We have dwelt considerably on the distinctive peculiarities of each of these engines, because we think it essential to place their respective attributes in strong relief before the apprehension of the public; and to define with clearness and accuracy the wholly different nature of the

* This subject is further noticed in Note F.

principles on which each is based, so as to make it self-evident to the reader (the mathematical reader at least) in what manner and degree the powers of the Analytical Engine transcend those of an engine, which, like the Difference Engine, can only work out such results as may be derived from *one restricted and particular series of processes*, such as those included in $\Delta^n u_z = 0$. We think this of importance, because we know that there exists considerable vagueness and inaccuracy in the mind of persons in general on the subject. There is a misty notion amongst most of those who have attended at all to it, that *two* 'calculating machines' have been successively invented by the same person within the last few years; whilst others again have never heard but of the one original 'calculating machine', and are not aware of there being any extension upon this. For either of these two classes of persons the above considerations are appropriate. Whilst the latter require a knowledge of the fact that there *are two* such inventions, the former are not less in want of accurate and well-defined information on the subject. No very clear or correct ideas prevail as to the characteristics of each engine, or their respective advantages or disadvantages; and, in meeting with those incidental allusions, of a more or less direct kind, which occur in so many publications of the day, to these machines, it must frequently be matter of doubt *which* 'calculating machine' is referred to, or whether *both* are included in the general allusion.

We are desirous likewise of removing two misapprehensions which we know obtain, to some extent, respecting these engines. In the first place it is very generally supposed that the Difference Engine, after it had been completed up to a certain point, *suggested* the idea of the Analytical Engine; and that the second is in fact the improved offspring of the first, and *grew out* of the existence of its predecessor, through some natural or else accidental combination of ideas suggested by this one. Such a supposition is in this instance contrary to the facts; although it seems to be almost an obvious inference, wherever two inventions, similar in their nature and objects, succeed each other closely in order of *time*, and strikingly in order of *value*; more especially / when the same individual is the author of both. Nevertheless the ideas which led to the Analytical Engine occurred in a manner wholly independent of any that were connected with the Difference Engine. These ideas are indeed in their own intrinsic nature independent of the latter engine, and might equally have occurred had it never existed nor been even thought of at all.

The second of the misapprehensions above alluded to, relates to the well-known suspension, during some years past, of all progress in the construction of the Difference Engine. Respecting the circumstances which have interfered with the actual completion of either invention, we offer no opinion; and in fact are not possessed of the data for doing so, had we the inclination. But we know that some persons suppose these obstacles (be they what they may) to have arisen *in consequence* of the subsequent invention of the Analytical Engine whilst the former was in progress. We have ourselves heard it even *lamented* that an idea should ever have occurred at all, which had turned out to be merely the means of arresting what was already in a course of successful execution, without substituting the superior invention in its stead. This notion we can contradict in the most unqualified manner. The progress of the Difference Engine had long been suspended, before there were even the least crude glimmerings of any invention superior to it. Such glimmerings, therefore, and their subsequent development, were in no way the original *cause* of that suspension; although, where difficulties of some kind or other evidently already existed, it was not perhaps calculated to remove or lessen them that an invention should have been meanwhile thought of, which, while including all that the first was capable of, possesses powers so extended as to eclipse it altogether.

We leave it for the decision of each individual (*after he has possessed himself* of competent information as to the characteristics of each engine), to determine how far it ought to be matter of regret that such an accession has been made to the powers of human science, even if it *has* (which we greatly doubt) increased to a certain limited extent some already existing difficulties that had arisen in the way of completing a valuable but lesser work. We leave it for each to satisfy himself as to the wisdom of desiring the obliteration (were that now possible) of all records of the more perfect invention, in order that the comparatively limited one might be finished. The Difference Engine would doubtless fulfil all those practical objects which it was originally destined for. It would certainly calculate all the tables that are more directly necessary for the physical purposes of life, such as nautical and other computations. Those who incline to very strictly utilitarian views, may perhaps feel that the peculiar powers of the Analytical Engine bear upon questions of abstract and speculative science, rather than upon those involving everyday and ordinary human interests. These persons being likely to possess but little sympathy, or possibly acquaintance, with any branches of science which they do not find to be *useful* (according to *their* definition of that word), may conceive that the undertaking of that engine, now that the other one is already in progress, would be a barren and unproductive laying out of yet more money and labour; in fact, a work of supererogation. Even in the utilitarian aspect, however, we do not doubt that very valuable practical results would be developed by the extended faculties of the Analytical Engine; some of which results / we think we could now hint at, had we the space; and others, which it may not yet be possible to foresee, but which would be brought forth by the daily increasing requirements of science, and by a more intimate practical acquaintance with the powers of the engine, were it in actual existence.

On general grounds, both of an *a priori* description as well as those founded on the scientific history and experience of mankind, we see strong presumptions that such would be the case. Nevertheless all will probably concur in feeling that the completion of the Difference Engine would be far preferable to the non-completion of any calculating engine at all. With whomsoever or wheresoever may rest the present causes of difficulty that apparently exist towards either the completion of the old engine, or the commencement of the new one, we trust they will not ultimately result in this generation's being acquainted with these inventions through the medium of pen, ink, and paper merely; and still more do we hope, that for the honour of our country's reputation in the future pages of history, these causes will not lead to the completion of the undertaking by some *other* nation or government. This could not but be matter of just regret; and equally so, whether the obstacles may have originated in private interests and feelings, in considerations of a more public description, or in causes combining the nature of both such solutions.

We refer the reader to the 'Edinburgh Review' of July 1834, for a very able account of the Difference Engine.[a] The writer of the article we allude to, has selected as his prominent matter for exposition, a wholly different view of the subject from that which M. Menabrea has chosen. The former chiefly treats it under its mechanical aspect, entering but slightly into the mathematical principles of which that engine is the representative, but giving, in considerable length, many details of the mechanism and contrivances by means of which it tabulates the various orders of differences. M. Menabrea, on the contrary, exclusively develops the analytical view; taking it for granted that mechanism is

[a] D. Lardner, 'Babbage's calculating engine' (1834), reprinted in *Works of Babbage*, Vol. 2.

able to perform certain processes, but without attempting to explain *how*; and devoting his whole attention to explanations and illustrations of the manner in which analytical laws can be so arranged and combined as to bring every branch of that vast subject within the grasp of the assumed powers of mechanism. It is obvious that, in the invention of a calculating engine, these two branches of the subject are equally essential fields of investigation, and that on their mutual adjustment, one to the other, must depend all success. They must be made to meet each other, so that the weak points in the powers of either department may be compensated by the strong points in those of the other. They are indissolubly connected, though so different in their intrinsic nature that perhaps the same mind might not be likely to prove equally profound or successful in both. We know those who doubt whether the powers of mechanism will in practice prove adequate in all respects to the demands made upon them in the working of such complicated trains of machinery as those of the above engines, and who apprehend that unforeseen practical difficulties and disturbances will arise in the way of accuracy and of facility of operation. The Difference Engine, however, appears to us to be in a great measure an answer to these doubts. It is complete as far as it goes, and it does work with all the anticipated success. The Analytical Engine, far from being / more complicated, will in many respects be of simpler construction; and it is a remarkable circumstance attending it, that with very *simplified* means it is so much more powerful.

The article in the 'Edinburgh Review' was written some time previous to the occurrence of any ideas such as afterwards led to the invention of the Analytical Engine; and in the nature of the Difference Engine there is much less that would invite a writer to take exclusively, or even prominently, the mathematical view of it, than in that of the Analytical Engine; although mechanism has undoubtedly gone much further to meet mathematics, in the case of this engine, than of the former one. Some publication embracing the *mechanical* view of the Analytical Engine is a desideratum which we trust will be supplied before long.

Those who may have the patience to study a moderate quantity of rather dry details, will find ample compensation, after perusing the article of 1834, in the clearness with which a succinct view will have been attained of the various practical steps through which mechanism can accomplish certain processes; and they will also find themselves still further capable of appreciating M. Menabrea's more comprehensive and generalized memoir. The very difference in the style and object of these two articles, makes them peculiarly valuable to each other; at least for the purposes of those who really desire something more than a merely superficial and popular comprehension of the subject of calculating engines. A.A.L.

Note B. Page 99

That portion of the Analytical Engine here alluded to is called the storehouse. It contains an indefinite number of the columns of discs described by M. Menabrea. The reader may picture to himself a pile of rather large draughtsmen heaped perpendicularly one above another to a considerable height, each counter having the digits from 0 to 9 inscribed on its *edge* at equal intervals; and if he then conceives that the counters do not actually lie one upon another so as to be in contact, but are fixed at small intervals of vertical distance on a common axis which passes perpendicularly through their centres, and around which each disc can *revolve horizontally* so that any required digit amongst those inscribed on its margin can be brought into view, he will have a good idea of one of these columns. The *lowest* of the discs on any column belongs to the units, the next above to the tens, the next above this to the hundreds, and so on. Thus, if we wished to inscribe 1345 on a column of the engine, it would stand thus:

1
3
4
5

In the Difference Engine there are seven of these columns placed side by side in a row, and the working mechanism extends behind them; the general form of the whole mass of machinery is that of a quadrangular prism (more or less approaching to the cube); the results always appearing on that perpendicular face of the engine which contains the columns of discs, opposite to which face a spectator may place himself. In the Analytical Engine there would be many more of these columns, probably at least two hundred. The precise form and arrangement / which the whole mass of its mechanism will assume is not yet finally determined.

We may conveniently represent the columns of discs on paper in a diagram like the following:

V_1	V_2	V_3	V_4	etc.
○	○	○	○	etc.
0	0	0	0	
0	0	0	0	
0	0	0	0	etc.
0	0	0	0	
□	□	□	□	etc.

The V's are for the purpose of convenient reference to any column, either in writing or speaking, and are consequently numbered. The reason why the letter V is chosen for this purpose in preference to any other letter, is because these columns are designated (as the reader will find in proceeding with the memoir) the *Variables*, and sometimes the *Variable columns*, or the *columns of Variables*. The origin of this appellation is, that the values on the columns are destined to change, that is to *vary*, in every conceivable manner. But it is necessary to guard against the natural misapprehension that the columns are only intended to receive the values of the *variables* in an analytical formula, and not of the *constants*. The columns are called Variables on a ground wholly unconnected with the *analytical* distinction between constants and variables. In order to prevent the possibility of confusion, we have, both in the translation and in the notes, written Variable with a capital letter when we use the word to signify a *column of the engine*, and variable with a small letter when we mean the *variable of a formula*. Similarly, *Variable cards* signify any cards that belong to a column of the engine.

To return to the explanation of the diagram: each circle at the top is intended to contain the algebraic sign + or −, either of which can be substituted* for the other, according as the number represented on the column below is positive or negative. In a similar manner any other

* A fuller account of the manner in which the *signs* are regulated, is given in M. Menabrea's Memoir, pages, 105, 106. He himself expresses doubts (in a note of his own at the bottom of the latter page) as to his having been likely to hit on the precise methods really adopted; his explanation being merely a conjectural one. That it *does* accord precisely with the fact is a remarkable circumstance, and affords a convincing proof how completely M. Menabrea has been imbued with the true spirit of the invention. Indeed the whole of the above Memoir is a striking production, when we consider that M. Menabrea had had but very slight means for obtaining any adequate ideas respecting the Analytical Engine. It requires, however, a considerable acquaintance with the abstruse and complicated nature of such a subject, in order fully to appreciate the penetration of the writer who could take so just and comprehensive a view of it upon such limited opportunity.

purely *symbolical* results of algebraical processes might be made to appear in these circles. In Note A, the practicability of developing *symbolical* with no less ease than *numerical* results has been touched on.

The zeros beneath the *symbolic* circles represent each of them a disc, supposed to have the digit 0 presented in front. Only four tiers of zeros have been figured in the diagram, but these may be considered as representing thirty or forty, or any number of tiers of discs that may be required. Since each disc can present any digit, and each circle any sign, the discs of every column may be so adjusted* as to express any positive or negative number whatever within the limits of the machine; which limits depend on the *perpendicular* extent of the mechanism, that is, on the number of discs to a column. /

Each of the squares below the zeros is intended for the inscription of any *general* symbol or combination of symbols we please; it being understood that the number represented on the column immediately above, is the numerical value of that symbol, or combination of symbols. Let us, for instance, represent the three quantities a, n, x, and let us further suppose that $a = 5$, $n = 7$, $x = 98$. We should have

V_1	V_2	V_3	V_4	etc.
+†	+	+	+	
0	0	0	0	
0	0	0	0	etc.
0	0	9	0	
5	7	8	0	etc.
[a]	[n]	[x]	[ ]	

We may now combine these symbols in a variety of ways, so as to form any required function or functions of them, and we may then inscribe each such function below brackets, every bracket uniting together those quantities (and those only) which enter into the function inscribed below it. We must also, when we have decided on the particular function whose numerical value we desire to calculate, assign another column to the right hand for receiving the *results*, and must inscribe the function in the square below this column. In the above instance we might have any one of the following functions:

$$ax^n,\ x^{an},\ a\cdot n\cdot x,\ \frac{a}{n}x,\ a + n + x, \text{ etc., etc.}$$

* This adjustment is done by hand merely.

† It is convenient to omit the circles whenever the signs + or − can be actually represented.

Let us select the first. It would stand as follows, previous to calculation:

V_1	V_2	V_3	V_4	etc.
+	+	+	+	
0	0	0	0	etc.
0	0	0	0	
0	0	9	0	
5	7	8	0	etc.
a	n	x	ax^n	etc.

$\underbrace{\boxed{a}\quad\boxed{n}\quad\boxed{x}}_{ax_n}$

The data being given, we must now put into the engine the cards proper for directing the operations in the case of the particular function chosen. These operations would in this instance be:

Firstly, six multiplications in order to get x^n $(=98^7$ for the above particular data).

Secondly, one multiplication in order then to get $a \cdot x^n$ $(=5 \cdot 98^7)$.

In all, seven multiplications to complete the whole process. We may thus represent them:

$$(\times, \times, \times, \times, \times, \times, \times), \text{ or } 7\,(\times).$$

The multiplications would, however, at successive stages in the solution of the problem, operate on pairs of numbers, derived from *different* columns. In other words, the *same operation* would be performed / on different *subjects of operation*. And here again is an illustration of the remarks made in the preceding note on the independent manner in which the engine directs its *operations*. In determining the value of ax^n, the *operations* are *homogeneous*, but are distributed amongst different *subjects of operation*, at successive stages of the computation. It is by means of certain punched cards, belonging to the Variables themselves, that the action of the operations is so *distributed* as to suit each particular function. The *operation cards* merely determine the succession of operations in a general manner. They in fact throw all that portion of the mechanism included in the *mill*, into a series of different *states*, which we may call the *adding state*, or the *multiplying state*, etc., respectively. In each of these states the mechanism is ready to act in the way peculiar to that state, on any pair of numbers which may be permitted to come within its sphere of action. Only *one* of these operating states of the mill can exist at a time; and the nature of the mechanism is also such that only *one pair of numbers* can be received and acted on at a time. Now, in order to secure that the mill shall receive a constant supply of the proper pairs of numbers in succession, and that it shall also rightly locate the result of an operation performed upon any pair, each Variable has cards of its own belonging to it. It has, first, a class of cards whose business it is to *allow* the number on the variable to pass into the mill, there to be operated upon. These cards may be called the *supplying cards. They* furnish the mill with its proper food. Each Variable has, secondly, another class of cards, whose office it is to allow the Variable to *receive* a number *from* the mill. These cards may be called the *receiving cards. They* regulate the location of results, whether temporary or ultimate results. The Variable cards in general (including both the preceding classes) might, it appears to us, be even more appropriately designated the distributive cards, since it is through their means that the action of the operations, and the results of this action, are rightly *distributed.*

There are *two varieties* of the *supplying* Variable cards, respectively adapted for fulfilling two distinct subsidiary purposes: but as these modifications do not bear upon the present subject, we shall notice them in another place.

In the above case of ax^n, the operation cards merely order seven multiplications, that is, they order the mill to be in the *multiplying state* seven successive times (without any reference to the particular columns whose numbers are to be acted upon). The proper distributive Variable cards step in at each successive multiplication, and cause the distributions requisite for the particular case.

For x^{an}	the operations would be	$34\ (\times)$
$a \cdot n \cdot x$		$(\times,\ \times)$, or $2\ (\times)$
$\frac{a}{n} \cdot x$		$(\div,\ \times)$
$a + n + x$		$(+,\ +)$, or $2\ (+)$

The engine might be made to calculate all these in succession. Having completed ax^n, the function x^{an} might be written under the brackets instead of ax^n, and a new calculation commenced (the appropriate operation and Variable cards for the new function of course coming into play). The results would then appear on V_5. So on for any number of different functions of the quantities *a*, *n*, *x*. Each *result* might / either permanently remain on its column during the succeeding calculations, so that when all the functions had been computed, their values would simultaneously exist on V_4, V_5, V_6, etc.; or each result might (after being printed off, or used in any specified manner) be effaced, to make way for its successor. The square under V_4 ought, for the latter arrangement, to have the functions ax^n, x^{an}, anx, etc., successively inscribed in it.

Let us now suppose that we have *two* expressions whose values have been computed by the engine independently of each other (each having its own group of columns for data and results). Let them be ax^n, $b \cdot p \cdot y$. They would then stand as follows on the columns:

V_1	V_2	V_3	V_4	V_5	V_6	V_7	V_8	V_9
+	+	+	+	+	+	+	+	+
0	0	0	0	0	0	0	0	0
0	0	0	0	0	0	0	0	0
0	0	0	0	0	0	0	0	0
0	0	0	0	0	0	0	0	0
a	n	x	ax^n	b	p	y	bpy	$\frac{ax^n}{bpy}$

We may now desire to combine together these two *results*, in any manner we please; in which case it would only be necessary to have an additional card or cards, which should order the requisite operations to be performed with the numbers on the two result columns, V_4 and V_8, and the *result of these further operations* to appear on a new column, V_9. Say that we wish to divide ax^n by $b \cdot p \cdot y$. The numerical value of this division would then appear on the column V_9, beneath which we have inscribed

$$\frac{ax^n}{bpy}.$$

The whole series of operations from the beginning would be as follows (n being $= 7$):

$$\{7(\times), 2(\times), \div\}, \text{ or } \{9(\times), \div\}.$$

This example is introduced merely to show that we may, if we please, retain separately and permanently any *intermediate* results (like ax^n, $b \cdot p \cdot y$), which occur in the course of processes having an ulterior and more complicated result as their chief and final object (like $ax^n/(bpy)$).

Any group of columns may be considered as representing a *general* function, until a *special* one has been implicitly impressed upon them through the introduction into the engine of the operation and Variable cards made out for a *particular* function. Thus, in the preceding example, V_1, V_2, V_3, V_5, V_6, V_7 represent the *general* function $\phi(a, n, b, p, x, y)$ until the function

$$\frac{ax^n}{b \cdot p \cdot y}$$

has been determined on, and *implicitly* expressed by the placing of the right cards in the engine. The actual working of the mechanism, as regulated by these cards, then *explicitly* develops the value of the function. The inscription of a function under the brackets, and in the square under the result column, in no way influences the processes or the results, and is merely a memorandum for the observer, to remind him of what is going on. It is the operation and the Variable cards only, which in reality determine the function. Indeed it should be distinctly kept in / mind that the inscriptions within *any* of the squares, are quite independent of the mechanism or workings of the engine, and are nothing but arbitrary memorandums placed there at pleasure to assist the spectator.

The further we analyse the manner in which such an engine performs its processes and attains its results, the more we perceive how distinctly it places in a true and just light the mutual relations and connection of the various steps of mathematical analysis, how clearly it separates those things which are in reality distinct and independent, and unites those which are mutually dependent. A.A.L.

Note C. Page 100

Those who may desire to study the principles of the Jacquard loom in the most effectual manner, viz. that of practical observation, have only to step into the Adelaide Gallery or the Polytechnic Institution. In each of these valuable repositories of scientific *illustration*, a weaver is constantly working at a Jacquard loom, and is ready to give any information that may be desired as to the construction and modes of acting of his apparatus. The volume on the manufacture of silk, in Lardner's Cyclopaedia, contains a chapter on the Jacquard loom, which may also be consulted with advantage.[a]

The mode of application of the cards, as hitherto used in the art of weaving, was not found, however, to be sufficiently powerful for all the simplifications which it was desirable to attain in such varied and

[a] D. Lardner, *The cabinet cyclopaedia* (133 vol., London, 1830–49).

complicated processes as those required in order to fulfil the purposes of an Analytical Engine. A method was devised of what was technically designated *backing* the cards in certain groups according to certain laws. The object of this extension is to secure the possibility of bringing any particular card or set of cards into use *any number of times successively* in the solution of one problem. Whether this power shall be taken advantage of or not, in each particular instance, will depend on the nature of the operations which the problem under consideration may require. The process is alluded to by M. Menabrea in pages 103, 104 and it is a very important simplification. It has been proposed to use it for the reciprocal benefit of that art, which, whilst it has itself no apparent connection with the domains of abstract science, has yet proved so valuable to the latter, in suggesting the principles which, in their new and singular field of application, seem likely to place *algebraical* combinations not less completely within the province of mechanism, than are all those varied intricacies of which *intersecting threads* are susceptible. By the introduction of the system of *backing* into the Jacquard loom itself, patterns which should possess symmetry, and follow regular laws of any extent, might be woven by means of comparatively few cards.

Those who understand the mechanism of this loom will perceive that the above improvement is easily effected in practice, by causing the prism over which the train of pattern cards is suspended, to revolve *backwards* instead of *forwards*, at pleasure, under the requisite circumstances; until, by so doing, any particular card, or set of cards, that has done duty once, and passed on in the ordinary regular succession, is brought back to the position it occupied just before it was used the preceding time. The prism then resumes its *forward* rotation, and thus brings the card or set of cards in question into play a second time. This process may obviously be repeated any number of times. A.A.L. /

Note D. Page 104

We have represented the solution of these two equations, with every detail, in a diagram* similar to those used in Note B; but additional explanations are requisite, partly in order to make this more complicated case perfectly clear, and partly for the comprehension of certain indications and notations not used in the preceding diagrams. Those

* See the diagram on page 141.

who may wish to understand Note G completely, are recommended to pay particular attention to the contents of the present note, or they will not otherwise comprehend the similar notation and indications when applied to a much more complicated case.

In all calculations, the columns of Variables used may be divided into three classes:

First. Those on which the data are inscribed:

Secondly. Those intended to receive the final results:

Thirdly. Those intended to receive such intermediate and temporary combinations of the primitive data as are not to be permanently retained, but are merely needed for *working with*, in order to attain the ultimate results. Combinations of this kind might properly be called *secondary data*. They are in fact so many *successive stages* towards the final result. The columns which receive them are rightly named *working Variables*, for their office is in its nature purely *subsidiary* to other purposes. They develop an intermediate and transient class of results, which unite the original data with the final results.

The result Variables sometimes partake of the nature of working Variables. It frequently happens that a Variable destined to receive a final result is the recipient of one or more intermediate values successively, in the course of the processes. Similarly, the Variables for data often become working Variables, or result Variables, or even both in succession. It so happens, however, that in the case of the present equations the three sets of offices remain throughout perfectly separate and independent.

It will be observed, that in the squares below the *working* Variables nothing is inscribed. Any one of these Variables is in many cases destined to pass through various values successively during the performance of a calculation (although in these particular equations no instance of this occurs). Consequently no *one fixed* symbol, or combination of symbols, should be considered as properly belonging to a merely *working* Variable; and as a general rule their squares are left blank. Of course in this, as in all other cases where we mention a *general* rule, it is understood that many particular exceptions may be expedient.

In order that all the indications contained in the diagram may be completely understood, we shall now explain two or three points, not hitherto touched on. When the value on any Variable is called into use, one of two consequences may be made to result. Either the value may *return* to the Variable after it has been used, in which case it is ready for a second use if needed; or the Variable may be made zero. (We are of course not considering a third case, of not unfrequent occurrence, in which the same Variable is destined to receive the *result* of the very operation which it has just supplied with a number.) Now the ordinary rule is, that the value *returns* to the Variable; unless it has been foreseen that no use for that value can recur, in which case / zero is substituted. At the *end* of a calculation, therefore, every column ought as a general rule to be zero, excepting those for results. Thus it will be seen by the diagram, that when *m*, the value on V_0, is used for the second time by operation 5, V_0 becomes 0, since *m* is not again needed; that similarly, when $(mn' - m'n)$, on V_{12}, is used for the third time by operation 11, V_{12} becomes zero, since $(mn' - m'n)$ is not again needed. In order to provide for the one or the other of the courses above indicated, there are *two* varieties of the *supplying* Variable cards. One of these varieties has provisions which cause the number given off from any Variable to *return* to that Variable after doing its duty in the mill. The other variety has provisions which cause *zero* to be substituted on the Variable, for the number given off. These two varieties are distinguished, when needful, by the respective appellations of the *retaining* supply cards and the *zero* supply cards. We see that the *primary* office (see Note B) of both these varieties of cards is the same; they only differ in their *secondary* office.

Every Variable thus has belonging to it *one* class of *receiving* Variable cards and *two* classes of *supplying* Variable cards. It is plain however that only the *one* or the *other* of these two latter classes can be used by any one Variable for *one* operation; never *both* simultaneously; their respective functions being mutually incompatible.

It should be understood that the Variable cards are not placed in *immediate contiguity* with the columns. Each card is connected by means of wires with the column it is intended to act upon.

Our diagram ought in reality to be placed side by side with M. Menabrea's corresponding table, so as to be compared with it, line for line belonging to each operation. But it was unfortunately inconvenient to print them in this desirable form. The diagram is, in the main, merely another manner of indicating the various relations denoted in M. Menabrea's table. Each mode has some advantages and some disadvantages. Combined, they form a complete and accurate method of registering every step and sequence in all calculations performed by the engine.

No notice has yet been taken of the *upper* indices which are added to the left of each *V* in the diagram; an addition which we have also taken the liberty of making to the *V*'s in M. Menabrea's tables of pages 104,

108, since it does not *alter* anything therein represented by him, but merely *adds* something to the previous indications of those tables. The *lower* indices are obviously indices of *locality* only, and are wholly independent of the operations performed or of the results obtained, their value continuing unchanged during the performance of calculations. The *upper* indices, however, are of a different nature. Their office is to indicate any *alteration* in the value which a Variable represents; and they are of course liable to changes during the processes of a calculation. Whenever a Variable has only zeros upon it, it is called ${}^{0}V$; the moment a value appears on it (whether that value be placed there arbitrarily, or appears in the natural course of a calculation), it becomes ${}^{1}V$. If this value gives place to another value, the Variable becomes ${}^{2}V$, and so forth. Whenever a *value* again gives place to zero, the Variable again becomes ${}^{0}V$, even if it had been ${}^{n}V$ the moment before. If a *value* then again be substituted, the Variable becomes ${}^{n+1}V$ (as it would have done if it had not passed through the intermediate ${}^{0}V$); etc., etc. Just before any calculation is commenced, and after the data have been given, and everything adjusted and prepared / for setting the mechanism in action, the upper indices of the Variables for data are all unity, and those for the working and result Variables are all zero. In this state the diagram represents them.*

There are several advantages in having a set of indices of this nature; but these advantages are perhaps hardly of a kind to be immediately perceived, unless by a mind somewhat accustomed to trace the successive steps by means of which the engine accomplishes its purposes. We have only space to mention in a general way, that the whole notation of the tables is made more consistent by these indices, for they are able to mark a *difference* in certain cases, where there would otherwise be an apparent *identity* confusing in its tendency. In such a case as $V_n = V_p + V_n$ there is more clearness and more consistency with the usual laws of algebraical notation, in being able to write ${}^{m+1}V_n = {}^{q}V_p + {}^{m}V_n$. It is also obvious that the indices furnish a powerful means of tracing back the derivation of any result; and of registering various circumstances concerning that *series of successive substitutions*, of which every *result* is in fact merely the final consequence; circum-

* We recommended the reader to trace the successive substitutions backwards from (1) to (4), in M. Menabrea's table. This he will easily do by means of the upper and lower indices, and it is interesting to observe how each V successively ramifies (so to speak) into two other V's in some other column of the table; until at length the V's of the original data are arrived at.

stances that may in certain cases involve relations which it is important to observe, either for purely analytical reasons, or for practically adapting the workings of the engine to their occurrence. The series of substitutions which lead to the equations of the diagram are as follows:

$$\begin{array}{ccccc} & (1) & (2) & (3) & (4) \\ {}^1V_{16}* = & \dfrac{{}^1V_{14}}{{}^1V_{12}} = & \dfrac{{}^1V_{10} - {}^1V_{11}}{{}^1V_6 - {}^1V_7} = & \dfrac{{}^1V_0 \cdot {}^1V_5 - {}^1V_2 \cdot {}^1V_3}{{}^1V_0 \cdot {}^1V_4 - {}^1V_3 \cdot {}^1V_1} = & \dfrac{d'm - dm'}{mn' - m'n} \end{array}$$

$$\begin{array}{ccccc} & (1) & (2) & (3) & (4) \\ {}^1V_{15} = & \dfrac{{}^1V_{13}}{{}^1V_{12}} = & \dfrac{{}^1V_8 - {}^1V_9}{{}^1V_6 - {}^1V_7} = & \dfrac{{}^1V_2 \cdot {}^1V_4 - {}^1V_5 \cdot {}^1V_1}{{}^1V_0 \cdot {}^1V_4 - {}^1V_3 \cdot {}^1V_1} = & \dfrac{dn' - d'n}{mn' - m'n} \end{array}$$

There are *three* successive substitutions for each of these equations. The formulae (2), (3), and (4) are *implicitly* contained in (1), which latter we may consider as being in fact the *condensed* expression of any of the former. It will be observed that every succeeding substitution must contain *twice* as many *V*'s as its predecessor. So that if a problem require *n* substitutions, the successive series of numbers for the *V*'s in the whole of them will be 2, 4, 8, 16 . . . 2^n.

The substitutions in the preceding equations happen to be of little value towards illustrating the power and uses of the upper indices; for owing to the nature of these particular equations the indices are all unity throughout. We wish we had space to enter more fully into the relations which these indices would in many cases enable us to trace.

M. Menabrea encloses the three centre columns of his table under the general title *Variable cards*. The *V*'s however in reality all represent the actual *Variable columns* of the engine, and not the cards that belong to them. Still the title is a very just one, since it is through the special action of certain Variable cards (when *combined* with the more generalized agency of the operation cards) that every one of the particular relations he has indicated under that title is brought about.

Suppose we wish to ascertain how often any *one* quantity, or combination / of quantities, is brought into use during a calculation. We easily ascertain *this*, from the inspection of any vertical column or columns of the diagram in which that quantity may appear. Thus, in the present case, we see that all the data, and all the intermediate results likewise, are used twice, excepting $(mn' - m'n)$, which is used three times.

The *order* in which it is possible to perform the operations for the present example, enables us to effect all the eleven operations of which it

* ⟨See footnote on page 138.⟩

consists, with only *three operation cards*; because the problem is of such a nature that it admits of each *class* of operations being performed in a group together; all the multiplications one after another, all the subtractions one after another, etc. The operations are

$$\{6(\times), 3(-), 2(\div)\}.$$

Since the very definition of an operation implies that there must be *two* numbers to act upon, there are of course *two supplying* Variable cards necessarily brought into action for every operation, in order to furnish the two proper numbers. (See Note B.) Also, since every operation must produce a *result*, which must be placed *somewhere*, each operation entails the action of a *receiving* Variable card, to indicate the proper locality for the result. Therefore, at least three times as many Variable cards as there are *operations* (not *operation cards*, for these, as we have just seen, are by no means always as numerous as the *operations*) are brought into use in every calculation. Indeed, under certain contingencies, a still larger proportion is requisite; such, for example, would probably be the case when the same result has to appear on more than one Variable simultaneously (which is not unfrequently a provision necessary for subsequent purposes in a calculation), and in some other cases which we shall not here specify. We see therefore that a great disproportion exists between the amount of *Variable* and of *operation* cards requisite for the working of even the simplest calculation.

All calculations do not admit, like this one, of the operations of the same nature being performed in groups together. Probably very few do so without exceptions occurring in one or other stage of the progress; and some would not admit it at all. The *order* in which the operations shall be performed in every particular case, is a very interesting and curious question, on which our space does not permit us fully to enter. In almost every computation a great *variety* of arrangements for the succession of the processes is possible, and various considerations must influence the selection amongst them for the purposes of a calculating engine. One essential object is to choose that arrangement which shall tend to reduce to a minimum the *time* necessary for completing the calculation.

It must be evident how multifarious and how mutually complicated are the considerations which the workings of such an engine involve. There are frequently several distinct *sets of effects* going on simultaneously; all in a manner independent of each other, and yet to a greater

DIAGRAM BELONGING TO NOTE D

Number of operations	Nature of operations	Variables for data						Working Variables									Variables for results	
		$^{1}V_{0}$ + 0 0 0 0	$^{1}V_{1}$ + 0 0 0 0	$^{1}V_{2}$ + 0 0 0 0	$^{1}V_{3}$ + 0 0 0 0	$^{1}V_{4}$ + 0 0 0 0	$^{1}V_{5}$ + 0 0 0 0	$^{0}V_{6}$ + 0 0 0 0	$^{0}V_{7}$ + 0 0 0 0	$^{0}V_{8}$ + 0 0 0 0	$^{0}V_{10}$ + 0 0 0 0	$^{0}V_{1}$ + 0 0 0 0	$^{0}V_{11}$ + 0 0 0 0	$^{0}V_{12}$ + 0 0 0 0	$^{0}V_{13}$ + 0 0 0 0	$^{0}V_{14}$ + 0 0 0 0	$^{0}V_{15}$ + 0 0 0 0	$^{0}V_{16}$ + 0 0 0 0
		m	n	d	m'	n'	d'										$\frac{dn' - d'n}{mn' - m'n} = x$	$\frac{d'm - dm'}{mn' - m'n} = y$
1	×	m	...	...	...	n'	...	mn'										
2	×	...	n	...	m'	...	...	...	$m'n$									
3	×	...	...	d	...	0	...	...	...	dn'								
4	×	...	0	...	...	...	d'	...	...	...	$d'n$							
5	×	0	...	...	...	...	0	...	...	...	...	$d'm$						
6	×	...	...	0	0	...	...	...	...	...	...	...	dm'					
7	−	...	...	...	...	...	...	0	0	...	...	...	...	$(mn' - m'n)$				
8	−	...	...	...	...	...	...	...	...	0	0	...	...	...	$(dn' - d'n)$			
9	−	...	...	...	...	...	...	...	...	...	...	0	0	...	...	$(d'm - dm')$		
10	÷	...	...	...	...	...	...	...	...	...	...	...	...	$(mn' - m'n)$	0	...	$\frac{dn' - d'n}{mn' - m'n} = x$	
11	÷	...	...	...	...	...	...	...	...	...	...	...	...	0	...	0	...	$\frac{d'm - dm'}{mn' - m'n} = y$

or less degree exercising a mutual influence. To adjust each to every other, and indeed even to perceive and trace them out with perfect correctness and success, entails difficulties whose nature partakes to a certain extent of those involved in every question where *conditions* are very numerous and inter-complicated; such as for instance the estimation of the mutual relations amongst *statistical* phenomena, and of those involved in many other classes of facts.

A.A.L. /

Note E. Page 107

This example has evidently been chosen on account of its brevity and simplicity, with a view merely to explain the *manner* in which the engine would proceed in the case of an *analytical calculation containing variables*, rather than to illustrate the *extent of its powers* to solve cases of a difficult and complex nature. The equations of page 102 are in fact a more complicated problem than the present one.

We have not subjoined any diagram of its development for this new example, as we did for the former one, because this is unnecessary after the full application already made of those diagrams to the illustration of M. Menabrea's excellent tables.

It may be remarked that a slight discrepancy exists between the formulae

$$(a + bx^1)$$

$$(A + B \cos^1 x)$$

given in the memoir as the *data* for calculation, and the *results* of the calculation as developed in the last division of the table which accompanies it. To agree perfectly with this latter, the data should have been given us

$$(ax^0 + bx^1)$$

$$(A \cos^0 x + B \cos^1 x).$$

The following is a more complicated example of the manner in which the engine would compute a trigonometrical function containing variables. To multiply

$$A + A_1 \cos \theta + A_2 \cos 2\theta + A_3 \cos 3\theta + \dots$$

by

$$B + B_1 \cos \theta.$$

Let the resulting products be represented under the general form

$$C_0 + C_1 \cos \theta + C_2 \cos 2\theta + C_3 \cos 3\theta + \ldots \qquad (1)$$

This trigonometrical series is not only in itself very appropriate for illustrating the processes of the engine, but is likewise of much practical interest from its frequent use in astronomical computations. Before proceeding further with it, we shall point out that there are three very distinct classes of ways in which it may be desired to deduce numerical values from any analytical formula.

First. We may wish to find the collective numerical value of the *whole formula*, without any reference to the quantities of which that formula is a function, or to the particular mode of their combination and distribution, of which the formula is the result and representative. Values of this kind are of a strictly arithmetical nature in the most limited sense of the term, and retain no trace whatever of the processes through which they have been deduced. In fact, any one such numerical value may have been attained from an *infinite variety* of data, or of problems. The values for x and y in the two equations (see Note D), come under this class of numerical results.

Secondly. We may propose to compute the collective numerical value of *each term* of a formula, or of a series, and to keep these results / separate. The engine must in such a case appropriate as many columns to *results* as there are terms to compute.

Thirdly. It may be desired to compute the numerical value of various *subdivisions of each term*, and to keep all these results separate. It may be required, for instance, to compute each coefficient separately from its variable, in which particular case the engine must appropriate *two* result columns to *every term that contains both a variable and coefficient*.

There are many ways in which it may be desired in special cases to distribute and keep separate the numerical values of different parts of an algebraical formula; and the power of effecting such distributions to any extent is essential to the *algebraical* character of the Analytical Engine. Many persons who are not conversant with mathematical studies, imagine that because the business of the engine is to give its results in *numerical notation*, the *nature of its processes* must consequently be *arithmetical* and *numerical*, rather than *algebraical* and *analytical.* This is an error. The engine can arrange and combine its numerical quantities exactly as if they were *letters* or any other *general* symbols; and in fact it might bring out its results in algebraical *notation*, were provisions made accordingly. It might develope three sets of results simultaneously, viz. *symbolic* results (as already alluded to in Notes A and B); *numerical* results (its chief and primary object); and *algebraical* results in *literal* notation. This latter however has not been deemed a necessary or desirable addition to its powers, partly because the necessary arrangements for effecting it would increase the complexity and extent of the mechanism to a degree that would not be commensurate with the advantages, where the main object of the invention is to translate into *numerical* language general formulae of analysis already known to us, or whose laws of formation are known to us. But it would be a mistake to suppose that because its *results* are given in the *notation* of a more restricted science, its *processes* are therefore restricted to those of that science. The object of the engine is in fact to give the *utmost practical efficiency* to the resources of *numerical interpretations* of the higher science of analysis, whilst it uses the processes and combinations of this latter.

To return to the trigonometrical series. We shall only consider the four first terms of the factor $(A + A_1 \cos\theta + \text{etc.})$, since this will be sufficient to show the method. We propose to obtain separately the numerical value of *each coefficient* C_0, C_1, etc., of (1). The direct multiplication of the two factors gives

$$\left.\begin{aligned} & BA + BA_1\cos\theta + BA_2\cos 2\theta + BA_3\cos 3\theta \\ & \quad + B_1A\cos\theta + B_1A_1\cos\theta\cdot\cos\theta + B_1A_2\cos 2\theta\cdot\cos\theta \\ & \qquad\qquad\qquad\qquad + B_1A_3\cos 3\theta\cdot\cos\theta \end{aligned}\right\} \quad (2)$$

a result which would stand thus on the engine:

Variables for data

V_0	V_1	V_2	V_3	...	V_{10}	V_{11}
A	A_1	A_2	A_3		B	B_1
	$\cos\theta$	$\cos 2\theta$	$\cos 3\theta$			$\cos\theta$

Variables for results

V_{20}	V_{21}	V_{22}	V_{23}	...	V_{31}	V_{32}	V_{33}	V_{34}
$\boxed{BA}$	$\boxed{BA_1}$	$\boxed{BA_2}$	$\boxed{BA_3}$		$\boxed{B_1A}$	$\boxed{B_1A_1}$	$\boxed{B_1A_2}$	$\boxed{B_1A_3}$
	$\cos\theta$	$\cos 2\theta$	$\cos 3\theta$		$\cos\theta$	$(\cos\theta\cdot\cos\theta)$	$(\cos 2\theta\cdot\cos\theta)$	$(\cos 3\theta\cdot\cos\theta)$

The variable belonging to each coefficient is written below it, as we have done in the diagram, by way of memorandum. The only further reduction which is at first apparently possible in the preceding result, would be the addition of V_{21} to V_{31} (in which case B_1A should be effaced from V_{31}). The whole operations from the beginning would then be:

First series of operations	*Second series of operations*	*Third series, which contains only one (final) operation*
${}^1V_{10}\times{}^1V_0={}^1V_{20}$	${}^1V_{11}\times{}^1V_0={}^1V_{31}$	${}^1V_{21}+{}^1V_{31}={}^2V_{21}$, *and*
${}^1V_{10}\times{}^1V_1={}^1V_{21}$	${}^1V_{11}\times{}^1V_1={}^1V_{32}$	V_{31} becomes $=0$
${}^1V_{10}\times{}^1V_2={}^1V_{22}$	${}^1V_{11}\times{}^1V_2={}^1V_{33}$	
${}^1V_{10}\times{}^1V_3={}^1V_{23}$	${}^1V_{11}\times{}^1V_3={}^1V_{34}$	

We do not enter into the same detail of *every* step of the processes as in the examples of Notes D and G, thinking it unnecessary and tedious to do so. The reader will remember the meaning and use of the upper and lower indices, etc., as before explained.

To proceed: we know that[a]

$$\cos n\theta\cdot\cos\theta=\tfrac{1}{2}\cos(n+1)\theta+\tfrac{1}{2}\cos(n-1)\theta \qquad (3)$$

Consequently, a slight examination of the second line of (2) will show that by making the proper substitutions, (2) will become

$$\begin{array}{r|l|l|l|l}
BA & +BA_1\ \cdot\cos\theta & +BA_2\ \cdot\cos 2\theta & +BA_3\ \cdot\cos 3\theta & \\
 & +B_1A\ \cdot\cos\theta & & & \\
+\tfrac{1}{2}B_1A_1 & & +\tfrac{1}{2}B_1A_1\cdot\cos 2\theta & & \\
 & +\tfrac{1}{2}B_1A_2\cdot\cos\theta & & +\tfrac{1}{2}B_1A_2\cdot\cos 3\theta & \\
 & & +\tfrac{1}{2}B_1A_3\cdot\cos 2\theta & & \\
 & & & & +\tfrac{1}{2}B_1A_3\cdot\cos 4\theta \\
C_0 & C_1 & C_2 & C_3 & C_4
\end{array}$$

[a] Equation corrected.

These coefficients should respectively appear on

V_{20} V_{21} V_{22} V_{23} V_{24}.

We shall perceive, if we inspect the particular arrangement of the results in (2) on the result columns as represented in the diagram, that, in order to effect this transformation, each successive coefficient upon V_{32}, V_{33}, etc. (beginning with V_{32}), must through means of proper cards be divided by *two*;* and that one of the halves thus / obtained must be added to the coefficient on the Variable which precedes it by ten columns, and the other half to the coefficient on the Variable which precedes it by twelve columns; V_{32}, V_{33}, etc. themselves becoming zeros during the process.

This series of operations may be thus expressed:

Fourth series

$$\begin{cases} {}^1V_{32} \div 2 + {}^1V_{22} = {}^2V_{22} = BA_2 + \frac{1}{2}B_1A_1 & \\ {}^1V_{32} \div 2 + {}^1V_{20} = {}^2V_{20} = BA + \frac{1}{2}B_1A_1 & = C_0 \end{cases}$$
$$\begin{cases} {}^1V_{33} \div 2 + {}^1V_{23} = {}^2V_{23} = BA_3 + \frac{1}{2}B_1A_2 & = C_3\dagger \\ {}^1V_{33} \div 2 + {}^2V_{21} = {}^3V_{21} = BA_1 + B_1A + \frac{1}{2}B_1A_2 \cdot\cdot & = C_1 \end{cases}$$
$$\begin{cases} {}^1V_{34} \div 2 + {}^0V_{24} = {}^1V_{24} = \frac{1}{2}B_1A_3 & = C_4 \\ {}^1V_{34} \div 2 + {}^2V_{22} = {}^3V_{22} = BA_2 + \frac{1}{2}B_1A_1 + \frac{1}{2}B_1A_3 & = C_2. \end{cases}$$

The calculation of the coefficients C_0, C_1, etc. of (1), would now be completed, and they would stand ranged in order on V_{20}, V_{21}, etc. It will be remarked, that from the moment the fourth series of operations is ordered, the Variables V_{31}, V_{32}, etc., cease to be *result* Variables, and become mere *working* Variables.

The substitution made by the engine of the processes in the second side of (3) for those in the first side, is an excellent illustration of the manner in which we may arbitrarily order it to substitute any function, number, or process, at pleasure, for any other function, number or process, on the occurrence of a specified contingency.

* This division would be managed by ordering the number two to appear on any separate new column which should be conveniently situated for the purpose, and then directing this column (which is in the strictest sense a *working* Variable) to divide itself successively with V_{32}, V_{33}, etc.

† It should be observed, that were the rest of the factor ($A + A\cos\theta +$, etc.) taken into account, instead of *four* terms only, C_3 would have the additional term $\frac{1}{2}B_1A_4$; and C_4 the two additional terms, BA_4, $\frac{1}{2}B_1A_5$. This would indeed have been the case had even *six* terms been multiplied.

We will now suppose that we desire to go a step further, and to obtain the numerical value of each *complete* term of the product (1), that is of each *coefficient and variable united*, which for the $(n + 1)$th term would be $C_n \cdot \cos n\theta$.

We must for this purpose place the variables themselves on another set of columns V_{41}, V_{42}, etc., and then order their successvie multiplication by V_{21}, V_{22}, etc., each for each. There would thus be a final series of operations as follows:

Fifth and final series of operations

$$^2V_{20} \times {}^0V_{40} = {}^1V_{40}$$
$$^3V_{21} \times {}^0V_{41} = {}^1V_{41}$$
$$^3V_{22} \times {}^0V_{42} = {}^1V_{42}$$
$$^2V_{23} \times {}^0V_{43} = {}^1V_{43}$$
$$^1V_{24} \times {}^0V_{44} = {}^1V_{44}$$

(N.B. that V_{40} being intended to receive the coefficient on V_{20} which has *no* variable, will only have $\cos 0\theta$ $(= 1)$ inscribed on it, preparatory to commencing the fifth series of operations.)

From the moment that the fifth and final series of operations is ordered, the Variables V_{20}, V_{21}, etc., then in their turn cease to be *result* / Variables and become mere *working* Variables; V_{40}, V_{41}, etc., being now the recipients of the ultimate results.

We should observe, that if the variables $\cos \theta$, $\cos 2\theta$, $\cos 3\theta$, etc., are furnished, they would be placed directly upon V_{41}, V_{42}, etc., like any other data. If not, a separate computation might be entered upon in a separate part of the engine, in order to calculate them, and place them on V_{41}, etc.

We have now explained how the engine might compute (1) in the most direct manner, supposing we knew nothing about the *general* term of the resulting series. But the engine would in reality set to work very differently, whenever (as in this case) we *do* know the law for the general term.

The two first terms of (1) are

$$(BA + \tfrac{1}{2}B_1A_1) + \{(BA_1 + B_1A + \tfrac{1}{2}B_1A_2) \cdot \cos \theta\} \qquad (4)$$

and the general term for all after these is

$$\{BA_n + \tfrac{1}{2}B_1 \cdot (A_{n-1} + A_{n+2})\} \cos n\theta \qquad (5)$$

which is the coefficient of the $(n + 1)$th term. The engine would calculate the two first terms by means of a separate set of suitable operation cards, and would then need another set for the third term; which last set of operation cards would calculate all the succeeding terms *ad infinitum*; merely requiring certain new Variable cards for each term to direct the operations to act on the proper columns. The following would be the successive sets of operations for computing the coefficients of $n + 2$ terms:

$$(\times, \times, \div, +), (\times, \times, \times, \div, +, +), n(\times, +, \times, \div, +).$$

Or we might represent them as follows, according to the numerical order of the operations:

$$(1, 2 \ldots 4), (5, 6 \ldots 10), n(11, 12 \ldots 15).$$

The brackets, it should be understood, point out the relation in which the operations may be *grouped*, whilst the comma marks *succession*. The symbol + might be used for this latter purpose, but this would be liable to produce confusion, as + is also necessarily used to represent one class of the actual operations which are the subject of that succession. In accordance with this meaning attached to the comma, care must be taken when any one group of operations recurs more than once, as is represented above by $n(11 \ldots 15)$, not to insert a comma after the number or letter prefixed to that group. $n, (11 \ldots 15)$ would stand for *an operation n, followed by the group of operations* $(11 \ldots 15)$; instead of denoting *the number of groups which are to follow each other*.

Wherever a *general term* exists, there will be a *recurring group* of operations, as in the above example. Both for brevity and for distinctness, a *recurring group* is called a *cycle*. A *cycle* of operations, then, must be understood to signify any *set of operations* which is repeated *more than once*. It is equally a *cycle*, whether it be repeated *twice* only, or an indefinite number of times; for it is the fact of a *repetition occurring at all* that constitutes it such. In many cases of / analysis there is a *recurring group* of one or more *cycles*; that is, a *cycle of a cycle*, or a *cycle of cycles*. For instance: suppose we wish to divide a series by a series,

$$\frac{a + bx + cx^2 + \ldots}{a' + b'x + c'x^2 + \ldots} \qquad (1)$$

it being required that the result shall be developed, like the dividend and the divisor, in successive powers of x. A little consideration of (1), and of the steps through which algebraical division is effected, will show that (if the denominator be supposed to consist of p terms) the first partial quotient will be completed by the following operations:

$$\{(\div), p(\times, -)\} \quad \text{or} \quad \{(1), p(2, 3)\} \tag{2}$$

that the second partial quotient will be completed by an exactly similar set of operations, which acts on the remainder obtained by the first set, instead of on the original dividend. The whole of the processes therefore that have been gone through, by the time the *second* partial quotient has been obtained, will be:

$$2\{(\div), p(\times, -)\} \quad \text{or} \quad 2\{(1), p(2, 3)\} \tag{3}$$

which is a cycle that includes a cycle, or a cycle of the second order. The operations for the *complete* division, supposing we propose to obtain n terms of the series constituting the quotient, will be:

$$n\{(\div), p(\times, -)\} \quad \text{or } n\{(1), p(2, 3)\}. \tag{4}$$

It is of course to be remembered that the process of algebraical division in reality continues *ad infinitum*, except in the few exceptional cases which admit of an exact quotient being obtained. The number n in the formula (4), is always that of the number of terms we propose to ourselves to obtain; and the nth partial quotient is the coefficient of the $(n - 1)$th power of x.

There are some cases which entail *cycles of cycles of cycles*, to an indefinite extent. Such cases are usually very complicated, and they are of extreme interest when considered with reference to the engine. The algebraical development in a series, of the nth function of any given function, is of this nature. Let it be proposed to obtain the nth function of

$$\phi(a, b, c \ldots x), \ x \text{ being the variable} \tag{5}$$

We should premise that we suppose the reader to understand what is meant by an nth function. We suppose him likewise to comprehend distinctly the difference between developing *an* nth *function algebraically*, and merely *calculating an* nth *function arithmetically*. If he does not, the following will be by no means very intelligible; but we have not space to give any preliminary explanations. To proceed: the law, according to which the successive functions of (5) are to be developed, must of course first be fixed on. This law may be of very various kinds. We may propose to obtain our results in successive *powers* of x, in which case the general form would be

$$C + C_1x + C_2x^2 + \ldots$$ /

or in successive powers of n itself, the index of the function we are ultimately to obtain, in which case the general form would be

$$C + C_1 n + C_2 n^2 + \ldots$$

and x would only enter in the coefficients. Again, other functions of x or of n instead of *powers*, might be selected. It might be in addition proposed, that the coefficients themselves should be arranged according to given functions of a certain quantity. Another mode would be to make equations arbitrarily amongst the coefficients only, in which case the several functions, according to either of which it might be possible to develop the nth function of (5), would have to be determined from the combined consideration of these equations and of (5) itself.

The *algebraical* nature of the engine (so strongly insisted on in a previous part of this note) would enable it to follow out any of these various modes indifferently; just as we recently showed that it can distribute and separate the numerical results of any one prescribed series of processes, in a perfectly arbitrary manner. Were it otherwise, the engine could merely *compute the arithmetical* nth *function*, a result which, like any other purely arithmetical results, would be simply a collective number, bearing no traces of the data or the processes which had led to it.

Secondly, the *law* of development for the nth function being selected, the next step would obviously be to develop (5) itself, according to this law. This result would be the first function, and would be obtained by a determinate series of processes. These in most cases would include amongst them one or more *cycles* of operations.

The third step (which would consist of the various processes necessary for effecting the actual substitution of the series constituting the *first function*, of the *variable* itself) might proceed in either of two ways. It might make the substitution either wherever x occurs in the original (5), or it might similarly make it wherever x occurs in the first function itself which is the equivalent of (5). In some cases the former mode might be best, and in others the latter.

Whichever is adopted, it must be understood that the result is to appear arranged in a series following the law originally prescribed for the development of the nth function. This result constitutes the second function; with which we are to proceed exactly as we did with the first function, in order to obtain the third function; and so on, $n - 1$ times, to obtain the nth function. We easily perceive that since every successive function is arranged in a series *following the same law*, there would (after the *first* function is obtained) be a *cycle, of a cycle, of a cycle*, etc., of operations,* one, two, three, up to $n - 1$ times, in order to get the nth function. We say, *after the first function is obtained*, because (for reasons on which we cannot here enter) the *first* / function might in many cases be developed through a set of processes peculiar to itself, and not recurring for the remaining functions.

We have given but a very slight sketch, of the principal *general* steps which would be requisite for obtaining an nth function of such a formula as (5). The question is so exceedingly complicated, that perhaps few persons can be expected to follow, to their own satisfaction, so brief and general a statement as we are here restricted to on this subject. Still it is a very important case as regards the engine, and suggests ideas peculiar to itself, which we should regret to pass wholly without allusion. Nothing could be more interesting than to follow out, in every detail, the solution by the engine of such a case as the above; but the time, space and labour this would necessitate, could only suit a very extensive work.

To return to the subject of *cycles* of operations: some of the notation of the integral calculus lends itself very aptly to express them: (2) might be thus written:

$$(\div), \Sigma(+1)^p (\times, -) \text{ or } (1), \Sigma(+1)^p (2, 3) \qquad (6)$$

where p stands for the variable; $(+1)^p$ for the function of the variable, that is, for ϕp; and the limits are from 1 to p, or from 0 to $p - 1$, each increment being equal to unity. Similarly, (4) would be:

$$\Sigma(+1)^n\{(\div), \Sigma(+1)^p (\times, -)\} \qquad (7)$$

the limits of n being from 1 to n, or from 0 to $n - 1$,

$$\text{or } \Sigma(+1)^n \{(1), \Sigma(+1)^p (2, 3)\}. \qquad (8)$$

Perhaps it may be thought that this notation is merely a circuitous way of expressing what was more simply and as effectually expressed before; and, in the above example, there may be some truth in this. But

* A cycle that includes n other cycles, successively *contained one within another*, is called a cycle of the $n + 1$th order. A cycle may simply *include* many other cycles, and yet only be of the second order. If a series follows a certain law for a certain number of terms, and then another law for another number of terms, there will be a cycle of operations for every new law; but these cycles will not be *contained one within another*, they merely *follow each other*. Therefore their number may be infinite without influencing the *order* of a cycle that includes a repetition of such a series.

there is another description of cycles which *can* only effectually be expressed, in a condensed form, by the preceding notation. We shall call them *varying cycles*. They are of frequent occurrence, and include successive cycles of operations of the following nature:

$$p(1, 2, \ldots m), (p-1)(1, 2, \ldots m),$$
$$(p-2)(1, 2, \ldots m) \ldots (p-n)(1, 2, \ldots m) \qquad (9)$$

where each cycle contains the same group of operations, but in which the number of repetitions of the group varies according to a fixed rate, with every cycle. (9) can be well expressed as follows:

$$\Sigma p(1, 2, \ldots m), \text{ the limits of } p \text{ being from } p-n \text{ to } p. \qquad (10)$$

Independent of the intrinsic advantages which we thus perceive to result in certain cases from this use of the notation of the integral calculus, there are likewise considerations which make it interesting, from the connections and relations involved in this new application. It has been observed in some of the former notes, that the processes used in analysis form a logical system of much higher generality than the applications to number merely. Thus, when we read over any algebraical formula, considering it exclusively with reference to the processes of the engine, and putting aside for the moment its abstract signification as to the relations of quantity, the symbols $+$, $\times$, etc., in reality represent (as their immediate and proximate effect, when the formula is / applied to the engine) that a certain prism which is a part of the mechanism (see Note C), turns a new face, and thus presents a new card to act on the bundles of levers of the engine; the new card being perforated with holes, which are arranged according to the peculiarities of the operation of addition, or of multiplication, etc. Again, the *numbers* in the preceding formula (8), each of them really represents one of these very pieces of card that are hung over the prism.

Now in the use made in the formulae (7), (8), and (10), of the notation of the integral calculus, we have glimpses of a similar new application of the language of the *higher* mathematics. Σ, in reality, here indicates that when a certain number of cards have acted in succession, the prism over which they revolve must *rotate backwards*, so as to bring those cards into their former position; and the limits 1 to n, 1 to p, etc., regulate how often this backward rotation is to be repeated.

A.A.L.

NOTE F. Page 111

There is in existence a beautiful woven portrait of Jacquard, in the fabrication of which 24,000 cards were required.

The power of *repeating* the cards, alluded to by M. Menabrea in page 103, and more fully explained in Note C, reduces to an immense extent the number of cards required. It is obvious that this mechanical improvement is especially applicable wherever *cycles* occur in the mathematical operations, and that, in preparing data for calculations by the engine, it is desirable to arrange the order and combination of the processes with a view to obtain them as much as possible *symmetrically* and in cycles, in order that the mechanical advantages of the *backing* system may be applied to the utmost. It is here interesting to observe the manner in which the value of an *analytical* resource is *met* and *enhanced* by an ingenious *mechanical* contrivance. We see in it an instance of one of those mutual *adjustments* between the purely mathematical and the mechanical departments, mentioned in Note A as being a main and essential condition of success in the invention of a calculating engine. The nature of the resources afforded by such adjustments would be of two principal kinds. In some cases, a difficulty (perhaps in itself insurmountable) in the one department, would be overcome by facilities in the other; and sometimes (as in the present case) a strong point in the one, would be rendered still stronger and more available, by combination with a corresponding strong point in the other.

As a mere example of the degree to which the combined systems of cycles and of backing can diminish the *number* of cards requisite, we shall choose a case which places it in strong evidence, and which has likewise the advantage of being a perfectly different *kind* of problem from those that are mentioned in any of the other notes. Suppose it be required to eliminate nine variables from ten simple equations of the form—

$$ax_0 + bx_1 + cx_2 + dx_3 + \ldots = p \qquad (1)$$

$$a^1x_0 + b^1x_1 + c^1x_2 + d^1x_3 + \ldots = p^1 \qquad (2)$$

$$\vdots$$

We should explain, before proceeding, that it is not our object to consider / this problem with reference to the actual arrangement of the data on the Variables of the engine, but simply as an abstract question of the *nature* and *number* of the *operations* required to be performed during its complete solution.

The first step would be the elimination of the first unknown quantity x_0 between the two first equations. This would be obtained by the form—

$$(a^1a - aa^1)x_0 + (a^1b - ab^1)x_1 + (a^1c - ac^1)x_2 + (a^1d - ad^1)x_3 + \ldots = a^1p - ap^1$$

for which the operations $10(\times, \times, -)$ would be needed. The second step would be the elimination of x_0 between the second and third equations, for which the operations would be precisely the same. We should then have had altogether the following operations—

$$10(\times, \times, -), 10(\times, \times, -), = 20(\times, \times, -).$$

Continuing in the same manner, the total number of operations for the complete elimination of x_0 between all the successive pairs of equations, would be—

$$9 \cdot 10(\times, \times, -) = 90(\times, \times, -).$$

We should then be left with nine simple equations of nine variables from which to eliminate the next variable x_1; for which the total of the processes would be—

$$8 \cdot 9(\times, \times, -) = 72(\times, \times, -).$$

We should then be left with eight simple equations of eight variables from which to eliminate x_2, for which the processes would be—

$$7 \cdot 8(\times, \times, -) = 56(\times, \times, -)$$

and so on. The total operations for the elimination of all the variables would thus be—

$$9 \cdot 10 + 8 \cdot 9 + 7 \cdot 8 + 6 \cdot 7 + 5 \cdot 6 + 4 \cdot 5 + 3 \cdot 4 + 2 \cdot 3 + 1 \cdot 2 = 330.$$

So that *three* operation cards would perform the office of 330 such cards.

If we take n simple equations containing $n - 1$ variables, n being a number unlimited in magnitude, the case becomes still more obvious, as the same three cards might then take the place of thousands or millions of cards.

We shall now draw further attention to the fact, already noticed, of its being by no means necessary that a formula proposed for solution should ever have been actually worked out, as a condition for enabling the engine to solve it. Provided we know the *series of operations* to be gone through, that is sufficient. In the foregoing instance this will be obvious enough on a slight consideration. And it is a circumstance which deserves particular notice, since herein may reside a latent value of such an engine almost incalculable in its possible ultimate results. We already know that there are functions whose numerical value it is of importance for the purposes both of abstract and of practical science to ascertain, but whose determination requires processes so lengthy and so complicated, that, although it is possible to arrive at them through great expenditure of time, labour, and money, it is yet on these accounts practically almost unattainable; and we can conceive there being some results which it may be *absolutely impossible* in practice to attain with any accuracy, and whose precise determination it may prove / highly important for some of the future wants of science in its manifold, complicated and rapidly developing fields of enquiry, to arrive at.

Without, however, stepping into the region of conjecture, we will mention a particular problem which occurs to us at this monent as being an apt illustration of the use to which such an engine may be turned for determining that which human brains find it difficult or impossible to work out unerringly. In the solution of the famous problem of the three bodies, there are, out of about 295 coefficients of lunar perturbations given by M. Clausen (Astro^{e}. Nachrichten, No. 406) as the result of the calculations by Burg, of two by Damoiseau, and of one by Burckhardt, fourteen coefficients that differ in the nature of their algebraic sign; and out of the remainder there are only 101 (or about one-third) that agree precisely both in signs and in amount. These discordances, which are generally small in individual magnitude, may arise either from an erroneous determination of the abstract coefficients in the development of the problem, or from discrepancies in the data deduced from observation, or from both causes combined. The former is the most ordinary source of error in astronomical computations, and this the engine would entirely obviate.

We might even invent laws for series or formulae in an arbitrary manner, and set the engine to work upon them, and thus deduce numerical results which we might not otherwise have thought of obtaining. But this would hardly perhaps in any instance be productive of any great practical utility, or calculated to rank higher than as a kind of philosophical amusement.

A.A.L.

NOTE G. Page 112

It is desirable to guard against the possibility of exaggerated ideas that might arise as to the powers of the Analytical Engine. In considering any new subject, there is frequently a tendency, first, to *overrate* what we find to be already interesting or remarkable; and, secondly, by a sort of natural reaction, to *undervalue* the true state of the case, when we do discover that out notions have surpassed those that were really tenable.

The Analytical Engine has no pretensions whatever to *originate* anything. It can do whatever we *know how to order it* to perform. It can *follow* analysis; but it has no power of *anticipating* any analytical relations or truths. Its province is to assist us in making *available* what we are already acquainted with. This it is calculated to effect primarily and chiefly of course, through its executive faculties; but it is likely to exert an *indirect* and reciprocal influence on science itself in another manner. For, in so distributing and combining the truths and the formulae of analysis, that they may become most easily and rapidly amenable to the mechanical combinations of the engine, the relations and the nature of many subjects in that science are necessarily thrown into new lights, and more profoundly investigated. This is a decidedly indirect, and a somewhat *speculative*, consequence of such an invention. It is however pretty evident, on general principles, that in devising for mathematical truths a new form in which to record and throw themselves out for actual use, views are likely to be induced, which should again react on the more theoretical phase of the subject. There are in all extensions of human power, or additions to human / knowledge, various *collateral* influences, besides the main and primary object attained.

To return to the executive faculties of this engine: the question must arise in every mind, are they *really* even able to *follow* analysis in its whole extent? No reply, entirely satisfactory to all minds, can be given to this query, excepting the actual existence of the engine, and actual experience of its practical results. We will however sum up for each reader's consideration the chief elements with which the engine works:

1. It performs the four operations of simple arithmetic upon any numbers whatever.

2. By means of certain artifices and arrangements (upon which we cannot enter within the restricted space which such a publication as the present may admit of), there is no limit either to the *magnitude* of the

numbers used, or to the *number* of *quantities* (either variables or constants) that may be employed.

3. It can combine these numbers and these quantities either algebraically or arithmetically, in relations unlimited as to variety, extent, or complexity.

4. It uses algebraic *signs* according to their proper laws, and develops the logical consequences of these laws.

5. It can arbitrarily substitute any formula for any other; effacing the first from the columns on which it is represented, and making the second appear in its stead.

6. It can provide for singular values. Its power of doing this is referred to in M. Menabrea's memoir, page 109, where he mentions the passage of values through zero and infinity. The practicability of causing it arbitrarily to change its processes at any moment, on the occurrence of any specified contingency (of which its substitution of $\{\frac{1}{2}\cos(n+1)\theta + \frac{1}{2}\cos(n-1)\theta\}$ for $(\cos n\theta \cdot \cos\theta)$ explained in Note E, is in some degree an illustration), at once secures this point.

The subject of integration and of differentiation demands some notice. The engine can effect these processes in either of two ways:

First. We may order it, by means of the operation and of the Variable cards, to go through the various steps by which the required *limit* can be worked out for whatever function is under consideration.

Secondly. It may (if we know the form of the limit for the function in question) effect the integration or differentiation by direct* substitution. / We remarked in Note B, that any *set* of columns on which numbers are inscribed, represents merely a *general* function of the several quantities, until the special function have been impressed by means of the operation and Variable cards. Consequently, if instead of requiring the value of the function, we require that of its integral, or of

* The engine cannot of course compute limits for perfectly *simple* and *uncompounded* functions, except in this manner. It is obvious that it has no power of representing or of manipulating with any but *finite* increments or decrements; and consequently that wherever the computation of limits (or of any other functions) depends upon the *direct* introduction of quantities which either increase or decrease *indefinitely*, we are absolutely beyond the sphere of its powers. Its nature and arrangements are remarkably adapted for taking into account all *finite* increments or decrements (however small or large), and for developing the true and logical modifications of form or value dependent upon differences of this nature. The engine may indeed be considered as including the whole calculus of finite differences; many of whose theorems would be especially and beautifully fitted for development by its processes, and would offer peculiarly interesting considerations. We may mention, as an example, the calculation of the numbers of Bernoulli by means of the *Differences of Nothing.*

its differential coefficient, we have merely to order whatever particular combination of the ingredient quantities may constitute that integral or that coefficient. In ax^n, for instance, instead of the quantities

$$\begin{array}{cccc} V_0 & V_1 & V_2 & V_3 \\ \boxed{a} & \boxed{n} & \boxed{x} & \boxed{ax^n} \end{array}$$

$$\underbrace{\boxed{a} \quad \boxed{n} \quad \boxed{x}}_{ax^n}$$

being ordered to appear on V_3 in the combination ax^n, they would be ordered to appear in that of

$$anx^{n-1}.$$

They would then stand thus:

$$\begin{array}{cccc} V_0 & V_1 & V_2 & V_3 \\ \boxed{a} & \boxed{n} & \boxed{x} & \boxed{anx^{n-1}} \end{array}$$

$$\underbrace{\boxed{a} \quad \boxed{n} \quad \boxed{x}}_{anx^{n-1}}$$

Similarly, we might have ax^{n+1}/n, the integral of ax^n.

An interesting example for following out the processes of the engine would be such a form as

$$\int \frac{x^n dx}{\sqrt{a^2 - x^2}}$$

or any other cases of integration by successive reductions, where an integral which contains an operation repeated n times can be made to depend upon another which contains the same $n - 1$ or $n - 2$ times, and so on until by continued reduction we arrive at a certain *ultimate* form, whose value has then to be determined.

The methods in Arbogast's *Calcul des Dérivations* are peculiarly fitted for the notation and the processes of the engine. Likewise the whole of the combinatorial analysis, which consists first in a purely numerical calculation of indices, and secondly in the distribution and combination of the quantities according to laws prescribed by these indices.

We will terminate these notes by following up in detail the steps through which the engine could compute the numbers of Bernoulli, this being (in the form in which we shall deduce it) a rather complicated example of its powers.[a] The simplest manner of computing these numbers would be from the direct expansion of

$$\frac{x}{e^x - 1} = \frac{1}{1 + \frac{x}{2} + \frac{x^2}{2\cdot 3} + \frac{x^3}{2\cdot 3\cdot 4} + \ldots} \qquad (1) \quad /$$

which is in fact a particular case of the development of

$$\frac{a + bx + cx^2 + \ldots}{a' + b'x + c'x^2 + \ldots}$$

mentioned in Note E. Or again, we might compute them from the well-known form

$$B_{2n-1} = 2\cdot\frac{1\cdot 2\cdot 3\cdot \ldots \cdot 2n}{(2\pi)^{2n}}\cdot\left\{1 + \frac{1}{2^{2n}} + \frac{1}{3^{2n}} + \ldots\right\} \qquad (2)$$

or from the form

$$B_{2n-1} = \frac{\pm 2n}{(2^{2n}-1)2^{n-1}}\left\{\begin{aligned}&\frac{1}{2}n^{2n-1}\\ &-(n-1)^{2n-1}\left\{1+\frac{1}{2}\cdot\frac{2n}{1}\right\}\\ &+(n-2)^{2n-1}\left\{1+\frac{2n}{1}+\frac{1}{2}\cdot\frac{2n\cdot(2n-1)}{1\cdot 2}\right\}\\ &-(n-3)^{2n-1}\left\{\begin{aligned}&1+\frac{2n}{1}+\frac{2n\cdot(2n-1)}{1\cdot 2}\\ &+\frac{1}{2}\cdot\frac{2n\cdot(2n-1)\cdot(2n-2)}{1\cdot 2\cdot 3}\end{aligned}\right\}\\ &+\ldots\end{aligned}\right\} \qquad (3)$$

[a] Lovelace's description of the Bernoulli numbers computation contains a number of minor errors, a state of affairs that computer practioners now regard as almost inevitable in unproven 'software'. Some corrections have been made in angle brackets. The 'diagram for the computation' for the Bernoulli numbers on pp. 168–9 contains an error: operation 24 should be a '−' rather than a '+'. Simulations by the editor, and independently by the Russian mathematicians A. E. and O. I. Petrenko, show that with this correction the process does indeed correctly compute Bernoulli numbers. (*Reference*: G. J. Tee in *Mathematical Reviews* (1983) 83b:01045.)

or from many others. As however our object is not simplicity or facility of computation, but the illustration of the powers of the engine, we prefer selecting the formula below, marked (8). This is derived in the following manner:

If in the equation

$$\frac{x}{e^x - 1} = 1 - \frac{x}{2} + B_1 \frac{x^2}{2} + B_3 \frac{x^4}{2 \cdot 3 \cdot 4} + B_5 \frac{x^6}{2 \cdot 3 \cdot 4 \cdot 5 \cdot 6} + \ldots \quad (4)$$

(in which B_1, B_3, etc., are the numbers of Bernoulli), we expand the denominator of the first side in powers of x, and then divide both numerator and denominator by x, we shall derive

$$1 = \left(1 - \frac{x}{2} + B_1 \frac{x^2}{2} + B_3 \frac{x^4}{2 \cdot 3 \cdot 4} + \ldots\right) \times \left(1 + \frac{x}{2} + \frac{x^2}{2 \cdot 3} + \frac{x^3}{2 \cdot 3 \cdot 4} \ldots\right) \quad (5)$$

If this latter multiplication be actually performed, we shall have a series of the general form

$$1 + D_1 x + D_2 x^2 + D_3 x^3 + \ldots \quad (6)$$

in which we see, first, that all the coefficients of the powers of x are severally equal to zero; and secondly, that the general form for D_{2n}, the coefficient of the $2n + 1$th *term*, (that is of x^{2n} any *even* power of x), is the following:

$$\left.\begin{aligned} &\frac{1}{2 \cdot 3 \cdot \ldots \cdot 2n+1} - \frac{1}{2} \cdot \frac{1}{2 \cdot 3 \cdot \ldots \cdot 2n} + \frac{B_1}{2} \cdot \frac{1}{2 \cdot 3 \cdot \ldots \cdot 2n-1} \\ &\quad + \frac{B_3}{2 \cdot 3 \cdot 4} \cdot \frac{1}{2 \cdot 3 \cdot \ldots \cdot 2n-3} + \frac{B_5}{2 \cdot 3 \cdot 4 \cdot 5 \cdot 6} \\ &\quad \times \frac{1}{2 \cdot 3 \cdot \ldots \cdot 2n-5} + \ldots + \frac{B_{2n-1}}{2 \cdot 3 \cdot \ldots \cdot 2n} \cdot 1 = 0 \end{aligned}\right\} \quad (7) \;/$$

Multiplying every term by $(2 \cdot 3 \cdot \ldots \cdot 2n)$, we have

$$\left.\begin{aligned} 0 = &-\frac{1}{2} \cdot \frac{2n-1}{2n+1} + B_1\left(\frac{2n}{2}\right) + B_3\left(\frac{2n \cdot 2n-1 \cdot 2n-2}{2 \cdot 3 \cdot 4}\right) \\ &+ B_5\left(\frac{2n \cdot 2n-1 \cdot \ldots \cdot 2n-4}{2 \cdot 3 \cdot 4 \cdot 5 \cdot 6}\right) + \ldots + B_{2n-1} \end{aligned}\right\} \quad (8)$$

which it may be convenient to write under the general form:

$$0 = A_0 + A_1B_1 + A_3B_3 + A_5B_5 + \ldots + B_{2n-1} \qquad (9)$$

A_1, A_3, etc., being those functions of n which respectively belong to B_1, B_3, etc.

We might have derived a form nearly similar to (8), from D_{2n-1} the coefficient of any *odd* power of x in (6); but the general form is a little different for the coefficients of the *odd* powers, and not quite so convenient.

On examining (7) and (8), we perceive that, when these formulae are isolated from (6) whence they are derived, and considered in themselves separately and independently, n may be any whole number whatever; although when (7) occurs *as one of the* D's in (6), it is obvious that n is then not arbitrary, but is always a certain function of the *distance of that* D *from the beginning*. If that distance be $= d$, then

$$2n + 1 = d, \text{ and } n = \frac{d-1}{2} \text{ (for any } even \text{ power of } x)$$

$$2n = d, \text{ and } n = \frac{d}{2} \text{ (for any } odd \text{ power of } x).$$

It is with the *independent* formula (8) that we have to do. Therefore it must be remembered that the conditions for the value of n are now modified, and that n is a perfectly *arbitrary* whole number. This circumstance, combined with the fact (which we may easily perceive) that whatever n is, every term of (8) after the $(n + 1)$th is $= 0$, and that the $(n + 1)$th term itself is always $= B_{2n-1} . (1/1) = B_{2n-1}$, enables us to find the value (either numerical or algebraical) of any nth number of Bernoulli B_{2n-1}, *in terms of all the preceding ones*, if we but know the values of $B_1, B_3, \ldots B_{2n-3}$. We append to this note a diagram and table, containing the details of the computation for B_7, (B_1, B_3, B_5 being supposed given).

On attentively considering (8), we shall likewise perceive that we may derive from it the numerical value of *every* number of Bernoulli in succession, from the very beginning, *ad infinitum*, by the following series of computations:

First series. Let $n = 1$, and calculate (8) for this value of n. The result is B_1.

Second series. Let $n = 2$. Calculate (8) for this value of n, substituting the value of B_1 just obtained. The result is B_3.

Third series. Let $n = 3$. Calculate (8) for this value of n, substituting the values of B_1, B_3 before obtained. The result is B_5. And so on, to any extent.

The diagram* represents the columns of the engine when just prepared / for computing B_{2n-1}, (in the case of $n = 4$); while the table beneath them presents a complete simultaneous view of all the successive changes which these columns then severally pass through in order to perform the computation. (The reader is referred to Note D, for explanations respecting the nature and notation of such tables.)

Six numerical *data* are in this case necessary for making the requisite combinations. These data are 1, 2, n (=4), B_1, B_3, B_5. Were $n = 5$, the additional datum B_7 would be needed. Were $n = 6$, the datum B_9 would be needed; and so on. Thus the actual *number of data* needed will always be $n + 2$, for $n = n$; and out of these $n + 2$ data, $(n + 2 - 3)$ of them are successive numbers of Bernoulli. The reason why the Bernoulli numbers used as data, are nevertheless placed on *result* columns in the diagram, is because they may properly be supposed to have been previously computed in succession by the *engine* itself; under which circumstances each B will appear as a *result*, previous to being used as a *datum* for computing the succeeding B. Here then is an instance (of the kind alluded to in Note D) of the same Variables filling more than one office in turn. It is true that if we consider our computation of B_7 as a perfectly *isolated* calculation, we may conclude B_1, B_3, B_5 to have been arbitrarily placed on the columns; and it would then perhaps be more consistent to put them on V_4, V_5, V_6 as data and not results. But we are not taking this view. On the contrary, we suppose the engine to be *in the course of* computing the numbers to an indefinite extent, from the very beginning; and that we merely single out, by way of example, *one amongst* the successive but distinct series of computations it is thus performing. Where the B's are fractional, it must be understood that they are computed and appear in the notation of *decimal* fractions. Indeed this is a circumstance that should be noticed with reference to all calculations. In any of the examples already given in the translation and in the notes, some of the *data*, or of the temporary or permanent results, might be fractional, quite as probably as whole numbers. But the arrangements are so made, that the nature of the processes would be the same as for whole numbers.

In the above table and diagram we are not considering the *signs* of

* See the diagram at the end of these notes ⟨pp. 168–9⟩.

any of the B's, merely their numerical magnitude. The engine would bring out the sign for each of them correctly of course, but we cannot enter on *every* additional detail of this kind, as we might wish to do. The circles for the signs are therefore intentionally left blank in the diagram.

Operation cards 1, 2, 3, 4, 5, 6 prepare

$$-\frac{1}{2}\cdot\frac{2n-1}{2n+1}.$$

Thus, Card 1 multiplies *two* into n, and the three *receiving* Variable cards belonging respectively to V_4, V_5, V_6, allow the result $2n$ to be placed on each of these latter columns (this being a case in which a triple receipt of the result is needed for subsequent purposes); we see that the upper indices of the two Variables used, during operation 1, remain unaltered.

We shall not go through the details of every operation singly, since the table and diagram sufficiently indicate them; we shall merely notice some few peculiar cases.

By operation 6, a *positive* quantity is turned into a *negative* quantity, by simply subtracting the quantity from a column which has only / zero upon it. (The sign at the top of $V_{\langle 13\rangle}$ would become – during this process).

Operation 7 will be unintelligible, unless it be remembered that if we were calculating for $n = 1$ instead of $n = 4$, operation 6 would have completed the computation of B_1 itself; in which case the engine, instead of continuing its processes, would have to put B_1 on V_{21}; and then either to stop altogether, or to begin operations 1, 2, . . . 7 all over again for value of n ($=2$), in order to enter on the computation of B_3; (having however taken care, previous to this recommencement, to make the number on V_3 equal to *two*, by the addition of unity to the former $n = 1$ on that column). Now operation 7 must either bring out a result equal to zero (if $n = 1$); or a result *greater* than *zero*, as in the present case; and the engine follows the one or the other of the two courses just explained, contingently on the one or the other result of operation 7. In order fully to perceive the necessity of this *experimental* operation, it is important to keep in mind what was pointed out, that we are not treating a perfectly isolated and independent computation, but one out of a series of antecedent and prospective computations.

Cards ⟨8, 9, 10, 11⟩ produce

$$-\frac{1}{2}\cdot\frac{2n-1}{2n+1}+B_1\frac{2n}{2}.$$

In operation 9 we see an example of an upper index which again becomes a value after having passed from preceding values to zero. V_{11} has successively been ${}^0V_{11}$, ${}^1V_{11}$, ${}^2V_{11}$, ${}^0V_{11}$, ${}^3V_{11}$; and, from the nature of the office which V_{11} performs in the calculation, its index will continue to go through further changes of the same description, which, if examined, will be found to be regular and periodic.

Card 12 has to perform the same office as card 7 did in the preceding section; since, if n had been $= 2$, the 11th operation would have completed the computation of B_3.

Cards 13 to 20 make A_3. Since A_{2n-1} always consists of $2n - 1$ factors, A_3 has three factors; and it will be seen that cards 13, 14, 15, 16 make the second of these factors, and then multiply it with the first; and that 17, 18, 19, 20 make the third factor, and then multiply this with the product of the two former factors.

Card 23 has the office of cards ⟨12⟩ and 7 to perform, since if n were $=$ 3, the 21st and 22nd operations would complete the computation of B_5. As our case is B_7, the computation will continue one more stage; and we must now direct attention to the fact, that in order to compute A_7 it is merely necessary precisely to repeat the group of operations 13 to 20; and then, in order to complete the computation of B_7, to repeat operations 21, 22.

It will be perceived that every unit added to n in B_{2n-1}, entails an additional repetition of operations (13 . . . 23) for the computation of B_{2n-1}. Not only are all the *operations* precisely the same however for every such repetition, but they require to be respectively supplied with numbers from the very *same pairs of columns*; with only the one exception of operation 21, which will of course need B_5 (from V_{23}) instead of B_3 (from V_{22}). This identity in the *columns* which supply the requisite numbers, must not be confounded with identity in the *values* those columns have upon them and give out to the mill. Most of those / values undergo alterations during a performance of the operations (13 . . . 23), and consequently the columns present a new set of values for the *next* performance of (13 . . . 23) to work on.

At the termination of the *repetition* of operations (13 . . . 23) in computing B_7, the alterations in the values on the Variables are, that

$V_6 = 2n - 4$ instead of $2n - 2$.
$V_7 = 6$ instead of 4.
$V_{10} = 0$ instead of 1.
$V_{13} = A_0 + A_1B_1 + A_3B_3 + A_5B_5$ instead of $A_0 + A_1B_1 + A_3B_3$

In this state the only remaining processes are first: to transfer the value which is on V_{13}, to V_{24}; and secondly to reduce V_6, V_7, V_{13} to zero, and to add* *one* to V_3, in order that the engine may be ready to commence computing B_9. Operations 24 and 25 accomplish these purposes. It may be thought anomalous that operation 25 is represented as leaving the upper index of V_3 still = unity. But it must be remembered that these indices always begin anew for a separate calculation, and that operation 25 places upon V_3 the *first* value *for the new calculation.*

It should be remarked, that when the group (13 ... 23) is *repeated*, changes occur in some of the *upper* indices during the course of the repetition: for example, ${}^3V_6$ would become ${}^4V_6$ and ${}^5V_6$.

We thus see that when $n = 1$, nine operation cards are used; that when $n = 2$, fourteen operation cards are used; and that when $n > 2$, twenty-five operation cards are used; but that no *more* are needed, however great n may be; and not only this, but that these same twenty-five cards suffice for the successive computation of all the numbers from B_1 to B_{2n-1} inclusive. With respect to the number of *Variable* cards, it will be remembered, from the explanations in previous notes, that an average of three such cards to each *operation* (not however to each operation card) is the estimate. According to this the computation of B_1 will require twenty-seven Variable cards; B_3 forty-two such cards; B_5 seventy-five; and for every succeeding B after B_5, there would be thirty-three additional Variable cards (since each repetition of the group (13 ... 23) adds eleven to the number of operations required for computing the previous B). But we must now explain, that whenever there is a *cycle of operations*, and if these merely require to be supplied with numbers from the *same pairs of columns* and likewise each operation to place its *result* on the *same* column for every repetition of the whole group, the process then admits of a *cycle of Variable cards* for effecting its purposes. There is obviously much more symmetry and simplicity in the arrangements, when cases do admit of repeating the Variable as well as the operation cards. Our present example is of this nature. The only exception to a *perfect identity* in *all* the processes and columns used, for every repetition of operations (13 ... 23) is, that operation 21 always requires one of its factors from a new column, and

* It is interesting to observe, that so complicated a case as this calculation of the Bernoullian numbers, nevertheless, presents a remarkable simplicity in one respect; viz., that during the processes for the computation of *millions* of these numbers, no other arbitrary modification would be requisite in the arrangements, excepting the above simple and uniform provision for causing one of the data periodically to receive the finite increment unity.

operation 24 always puts its result / on a new column. But as these variations follow the same law at each repetition (operation 21 always requiring its factor from a column *one* in advance of that which it used the previous time, and operation 24 always putting its result on the column *one* in advance of that which received the previous result) they are easily provided for in arranging the recurring group (or cycle) of Variable cards.

We may here remark that the average estimate of three Variable cards coming into use to each operation, is not to be taken as an absolutely and literally correct amount for all cases and circumstances. Many special circumstances, either in the nature of a problem, or in the arrangements of the engine under certain contingencies, influence and modify this average to a greater or less extent. But it is a very safe and correct *general* rule to go upon. In the preceding case it will give us seventy-five Variable cards as the total number which will be necessary for computing any B after B_3. This is very nearly the precise amount really used, but we cannot here enter into the minutiae of the few particular circumstances which occur in this example (as indeed at some one stage or other of probably most computations) to modify slightly this number.

It will be obvious that the very *same* seventy five Variable cards may be repeated for the computation of every succeeding number, just on the same principle as admits of the repetition of the thirty-three Variable cards of operations (13 . . . 23) in the computation of any *one* number. Thus there will be a *cycle of a cycle* of Variable cards.

If we now apply the notation for cycles, as explained in Note E, we may express the operations for computing the numbers of Bernoulli in the following manner:

(1 . . . 7), (24, 25) gives B_1 = first number; (n being = 1)

(1 . . . 7), (8 . . . 12), (24, 25) gives
B_3 = second number; (n being = 2)

(1 . . . 7), (8 . . . 12), (13 . . . 23), (24, 25) gives
B_5 = third number; (n being = 3)

(1 . . . 7), (8 . . . 12), 2(13 . . . 23), (24, 25) gives
B_7 = fourth number; (n being = 4)

⋮

(1 . . . 7), (8 . . . 12), $\Sigma(+1)^{n-2}$(13 . . . 23), (24, 25) gives
B_{2n-1} = nth number; (n being = n)

Again,

$$(1\ldots 7),\ (24,\ 25),$$

$$\underset{\text{limits 1 to } n}{\Sigma(+1)^n}\{(1\ldots 7),\ (8\ldots 12),\ \underset{\text{limits 0 to } (n+2)}{\Sigma(n+2)}(13\ldots 23),\ (24,\ 25)\}$$

represents the total operations for computing every number in succession, from B_1 to B_{2n-1} inclusive.

In this formula we see a *varying cycle* of the *first* order, and an ordinary cycle of the *second* order. The latter cycle in this case includes in it the varying cycle.

On inspecting the ten working Variables of the diagram, it will be perceived, that although the *value* on any one of them (excepting V_4 and V_5) goes through a series of changes, the *office* which each performs is in this calculation *fixed* and *invariable.* Thus V_6 always prepares the *numerators* of the factors of any A; V_7 the *denominators.* V_8 always receives the $(2n-3)$th factor of A_{2n-1}, and V_9 the $(2n-1)$th. V_{10} always decides which of two courses the succeeding processes are to follow, by feeling for the value of n through / means of a subtraction; and so on; but we shall not enumerate further. It is desirable in all calculations, so to arrange the processes, that the *offices* performed by the Variables may be as uniform and fixed as possible.

Supposing that it was desired not only to tabulate B_1, B_3, etc., but A_0, A_1, etc.; we have only then to appoint another series of Variables, V_{41}, V_{42}, etc., for receiving these latter results as they are successively produced upon V_{11}. Or again, we may, instead of this, or in addition to this second series of results, wish to tabulate the value of each successive *total* term of the series (8), viz: A_0, A_1B_1, A_3B_3, etc. We have then merely to multiply each B with each corresponding A, as produced; and to place these successive products on result columns appointed for the purpose.

The formula (8) is interesting in another point of view. It is one particular case of the general integral of the following equation of mixed differences:[a]

$$\frac{d^2}{dx^2}(z_{n+1}x^{2n+2}) = (2n+1)(2n+2)z^n x^{2n}$$

for certain special suppositions respecting z, x and n.

[a] *Sic* the equation.

Number of Operation.	Nature of Operation.	Variables acted upon.	Variables receiving results.	Indication of change in the value on any Variable.	Statement of Results.	Data.					
						$^1V_1$ ○ 0 0 1 [1]	$^1V_2$ ○ 0 0 2 [2]	$^1V_3$ ○ 0 0 4 [n]	$^0V_4$ ○ 0 0 0 [ ]	$^0V_5$ ○ 0 0 0 [ ]	
1	×	$^1V_2 \times {}^1V_3$	$^1V_4$, $^1V_5$, $^1V_6$	$^1V_2 = {}^1V_2$, $^1V_3 = {}^1V_3$	$= 2n$	...	2	n	$2n$	$2n$	
2	−	$^1V_4 - {}^1V_1$	$^2V_4$	$^1V_4 = {}^2V_4$, $^1V_1 = {}^1V_1$	$= 2n - 1$	1	...	...	$2n - 1$		
3	+	$^1V_5 + {}^1V_1$	$^2V_5$	$^1V_5 = {}^2V_5$, $^1V_1 = {}^1V_1$	$= 2n + 1$	1	...	...	...	$2n + 1$	
4	÷	$^2V_5 \div {}^2V_4$	$^1V_{11}$	$^2V_5 = {}^0V_5$, $^2V_4 = {}^0V_4$	$= \frac{2n-1}{2n+1}$	...	...	...	0	0	
5	÷	$^1V_{11} \div {}^1V_2$	$^2V_{11}$	$^1V_{11} = {}^2V_{11}$, $^1V_2 = {}^1V_2$	$= \frac{1}{2} \cdot \frac{2n-1}{2n+1}$	...	2	...	...	...	
6	−	$^0V_{13} - {}^2V_{11}$	$^1V_{13}$	$^2V_{11} = {}^0V_{11}$, $^0V_{13} = {}^1V_{13}$	$= -\frac{1}{2} \cdot \frac{2n-1}{2n+1} = A_0$	...	...	...	...	...	
7	−	$^1V_3 - {}^1V_1$	$^1V_{10}$	$^1V_3 = {}^1V_3$, $^1V_1 = {}^1V_1$	$= n - 1 \ (= 3)$	1	...	n	...	...	
8	+	$^1V_2 + {}^0V_7$	$^1V_7$	$^1V_2 = {}^1V_2$, $^0V_7 = {}^1V_7$	$= 2 + 0 = 2$	...	2	...	...	...	
9	÷	$^1V_6 \div {}^1V_7$	$^3V_{11}$	$^1V_6 = {}^1V_6$, $^0V_{11} = {}^3V_{11}$	$= \frac{2n}{2} = A_1$	...	...	...	...	...	
10	×	$^1V_{21} \times {}^3V_{11}$	$^1V_{12}$	$^1V_{21} = {}^1V_{21}$, $^3V_{11} = {}^3V_{11}$	$= B_1 \cdot \frac{2n}{2} = B_1 A_1$	...	...	...	...	...	
11	+	$^1V_{12} + {}^1V_{13}$	$^2V_{13}$	$^1V_{12} = {}^0V_{12}$, $^1V_{13} = {}^2V_{13}$	$= -\frac{1}{2} \cdot \frac{2n-1}{2n+1} + B_1 \cdot \frac{2n}{2}$ ...	...	...	...	...	...	
12	−	$^1V_{10} - {}^1V_1$	$^2V_{10}$	$^1V_{10} = {}^2V_{10}$, $^1V_1 = {}^1V_1$	$= n - 2 \ (= 2)$	1	...	...	...	...	
13	−	$^1V_6 - {}^1V_1$	$^2V_6$	$^1V_6 = {}^2V_6$, $^1V_1 = {}^1V_1$	$= 2n - 1$	1	...	...	...	...	$2n$
14	+	$^1V_1 + {}^1V_7$	$^2V_7$	$^1V_1 = {}^1V_1$, $^1V_7 = {}^2V_7$	$= 2 + 1 = 3$	1	...	...	...	...	
15	÷	$^2V_6 \div {}^2V_7$	$^1V_8$	$^2V_6 = {}^2V_6$, $^2V_7 = {}^2V_7$	$= \frac{2n-1}{3}$	...	...	...	...	...	$2n$
16	×	$^1V_8 \times {}^3V_{11}$	$^4V_{11}$	$^1V_8 = {}^0V_8$, $^3V_{11} = {}^4V_{11}$	$= \frac{2n}{2} \cdot \frac{2n-1}{3}$	...	...	...	...	...	
17	−	$^2V_6 - {}^1V_1$	$^3V_6$	$^2V_6 = {}^3V_6$, $^1V_1 = {}^1V_1$	$= 2n - 2$	1	...	...	...	...	$2n$
18	+	$^1V_1 + {}^2V_7$	$^3V_7$	$^2V_7 = {}^3V_7$, $^1V_1 = {}^1V_1$	$= 3 + 1 = 4$	1	...	...	...	...	
19	÷	$^3V_6 \div {}^3V_7$	$^1V_9$	$^3V_6 = {}^3V_6$, $^3V_7 = {}^3V_7$	$= \frac{2n-2}{4}$	...	...	...	...	...	$2n$
20	×	$^1V_9 \times {}^4V_{11}$	$^5V_{11}$	$^1V_9 = {}^0V_9$, $^4V_{11} = {}^5V_{11}$	$= \frac{2n}{2} \cdot \frac{2n-1}{3} \cdot \frac{2n-2}{4} = A_3$	...	...	...	...	...	
21	×	$^1V_{22} \times {}^5V_{11}$	$^0V_{12}$	$^1V_{22} = {}^1V_{22}$, $^0V_{12} = {}^2V_{12}$	$= B_3 \cdot \frac{2n}{2} \cdot \frac{2n-1}{3} \cdot \frac{2n-2}{3} = B_3 A_3$	...	...	...	...	...	
22	+	$^2V_{12} + {}^2V_{13}$	$^3V_{13}$	$^2V_{12} = {}^0V_{12}$, $^2V_{13} = {}^3V_{13}$	$= A_0 + B_1 A_1 + B_3 A_3$	...	...	...	...	...	
23	−	$^2V_{10} - {}^1V_1$	$^3V_{10}$	$^2V_{10} = {}^3V_{10}$, $^1V_1 = {}^1V_1$	$= n - 3 \ (= 1)$	1	...	...	...	...	
								Here follows a repeti			
24	+	$^4V_{13} + {}^0V_{24}$	$^1V_{24}$	$^4V_{13} = {}^0V_{13}$, $^0V_{24} = {}^1V_{24}$	$= B_7$	...	...	...	...	...	
25	+	$^1V_1 + {}^1V_3$	$^1V_3$	$^1V_1 = {}^1V_1$, $^1V_3 = {}^1V_3$, $^5V_6 = {}^0V_6$, $^5V_7 = {}^0V_7$	$= n + 1 = 4 + 1 = 5$ .... by a Variable-card. by a Variable card.	1	...	$n + 1$	...	...	

Diagram for the computation by the Engine of the Numbers of Bernoulli.

Working Variables.						Result Variables.			
$^0V_8$ ○ 0 0 0	$^0V_9$ ○ 0 0 0	$^0V_{10}$ ○ 0 0 0	$^0V_{11}$ ○ 0 0 0	$^0V_{12}$ ○ 0 0 0	$^0V_{13}$ ○ 0 0 0	$^1V_{21}$ B_1 in a decimal ○ fraction.	$^1V_{22}$ B_3 in a decimal ○ fraction.	$^1V_{23}$ B_5 in a decimal ○ fraction.	$^0V_{24}$... ○ 0 0 0
☐	☐	☐	☐	☐	☐	B_1	B_3	B_5	B_7
...	...	...	$\frac{2n-1}{2n+1}$						
...	...	...	$\frac{1}{2}\cdot\frac{2n-1}{2n+1}$						
...	...	...	0	....	$-\frac{1}{2}\cdot\frac{2n-1}{2n+1}=A_0$				
...	...	$n-1$							
...	...	...	$\frac{2n}{2}=A_1$						
...	...	...	$\frac{2n}{2}=A_1$	$B_1\cdot\frac{2n}{2}=B_1A_1$	....	B_1			
...	...	...	....	0	$\left\{-\frac{1}{2}\cdot\frac{2n-1}{2n+1}+B_1\cdot\frac{2n}{2}\right\}$				
...	...	$n-2$							
$\frac{2n-1}{3}$									
0	...	...	$\frac{2n}{2}\cdot\frac{2n-1}{3}$						
...	$\frac{2n-2}{4}$	...	$\left\{\frac{2n}{2}\cdot\frac{2n-1}{3}\cdot\frac{2n-2}{3}=A_3\right\}$						
...	0								
...	...	...	0	B_3A_3	....	......	B_3		
...	...	...	....	0	$\left\{A_3+B_1A_1+B_3A_3\right\}$				
...	...	$n-3$							
...tions thirteen to twenty-three.									
...	...	...	....	....	....	......	......	......	B_7

Diagram for the computation by the Engine of the Numbers of Bernoulli.

The *general* integral itself is of the form,

$$z_n = f(n) \cdot x + f_1(n) + f_2(n) \cdot x^{-1} + f_3(n) \cdot x^{-3} + \ldots$$

and it is worthy of remark, that the engine might (in a manner more or less similar to the preceding) calculate the value of this formula upon most *other* hypotheses for the functions in the integral, with as much, or (in many cases) with more, ease than it can formula (8).

A.L.L. /

VIII

LETTER TO THE EDITOR OF *COSMOS*[a]

BY L. F. MENABREA

From *Cosmos*, Vol. 6 (1855), pp. 421–2

L'insertion dans le *Cosmos* du discours du comte Rosee, où le noble Président de la Société Royale plaidait avec tant de chaleur la cause de la machine à calcul de M. Babbage, nous a valu l'honneur d'une lettre de M. Ménabréa, savant italien très-distingué, qui s'est aussi occupé de cette question. Voici la lettre du mathématicien turinais:

En parcourant la livraison du 6 avril courant de votre intéressant journal le *Cosmos*, j'ai appris avec un vif plaisir que l'on s'occupait sérieusement en Angleterre de mettre à exécution la *Machine analytique* de M. Charles Babbage. Dans le discours où M. le comte Rosse rend compte des démarches qui ont déjà été faites dans ce but, j'ai surtout remarqué le passage suivant: 'Avant de prendre une détermination et de me mettre en campagne, j'écrivis à plusieurs hommes éminents par leur savoir, et leur demandai si, dans leur opinion, on aurait fait un grand pas au point de vue de la science théorique et pratique, si les vues de M. Babbage, telles qu'elles sont exposées dans le petit essai publié par lui, sous le titre de *Ménabréa*, étaient complétement réalisées. Leurs réponses furent unanimement et fortement affirmatives.'

Comme je ne suis pas entièrement étranger à cette question, permettez-moi, Monsieur, de vous communiquer quelques détails qui ne seront pas sans intérêt pour les personnes qui s'occupent de science. Je suis bien réellement l'auteur du petit écrit que M. le comte Rosse attribue à M. Babbage, sous le pseudonyme de Ménabrea. Il y a bien des années, M. Babbage, lors d'un

[a] A new English translation of this letter appears as the next article in this volume.

voyage qu'il fit en Italie, s'arrêta quelque temps à Turin, où il eut la bonté de m'expliquer les dispositions principales de sa *machine analytique*, qui diffère essentiellement de la *machine aux différences* déjà connue du public. A dire vrai, le problème que s'était proposé / M. Babbage était tellement singulier, que mon premier sentiment fut celui du doute. Mais en y réfléchissant, je parvins à me convaincre que l'idée de cet illustre savant était parfaitement, rationnelle et, du consentement de l'auteur, je pris le parti de faire connaître les principes fondamentaux sur lesquels repose l'étonnant instrument dont il s'agit, dans un article qui fut publié dans la *Bibliothèque universelle de Genève*, n° 82, octobre 1842, page 352 et suivantes.

Quelque temps après, parut une traduction anglaise de ce même écrit, intitulée: *Sketch of the analytical Engine, invented by Charles Babbage Esq. by L. F. Ménabréa of Turin officer of the miltary Engineers*, with notes by translator. (*Extracted from the scientific memoirs, vol.* III. London, *printed by Richard and John E. Taylor – Reed Liancourt Fleet Street*, 1843.) Les notes qui accompagnaient la traduction de mon petit mémoire étaient extrêmement remarquables et annonçaient, dans leur auteur, une sagacité peu ordinaire. J'ignorais le nom de cet auteur lorsque, à mon grand étonnement, j'appris de M. Babbage lui-même que la traduction et les notes étaient l'ouvrage de lady Adda Lovelace, de la fille de Lord Byron, dame aussi distinguée par l'élévation de son esprit que remarquable par sa beauté, et que la mort a enlevée, il y a peu d'années, dans l'âge le plus brillant de la vie.

Je croyais cet écrit oublié; mais, puisqu'il vient de servir de point de départ aux démarches qui ont été faites auprès du gouvernement anglais pour réaliser l'invention de M. Babbage, j'ai cru devoir saisir cette occasion pour rendre hommage à la mémoire de la fille du grand poëte et attirer l'attention des savants, non pas sur mon ouvrage qui est bien modeste, mais sur les notes et commentaires qui en accompagnent la traduction et qui sont de nature à faire connaître le but et la puissance de la *Machine analytique*. Cette invention ne ressemble à rien de ce qui a été imaginé jusqu'à ce jour, et je ne pourrais, dans les limites d'une simple lettre, en faire connaître les principes; mais qu'il suffise de dire que, par son moyen, on pourrait exécuter la série des opérations analytiques et numériques qu'exige la solution d'un problème déterminé, de la même manière que, dans le métier à la *Jacquard*, par l'emploi des *cartons*, on exécute les dessins d'une étoffe brochée.

Agréez, Monsieur le directeur, l'expression de la considération très-distinguée avec laquelle j'ai l'honneur d'être.

L. F. MENABREA
Membre de l'Académie R. des sciences et du Parlement sarde /

IX

LETTER TO THE EDITOR OF *COSMOS*[a]

BY L. F. MENABREA

Translated from *Cosmos*, Vol. 6 (1855), pp. 421–2

The inclusion in *Cosmos* of Count Rosse's speech in which the noble President of the Royal Society pleaded the cause of Mr Babbage's calculating machine with such warmth, led M. Menabrea, the highly distiguished Italian scientist who has also been working on this matter, to honour us with a letter. Here is the Turin mathematician's letter:

While perusing the issue of 6 April current of your interesting journal *Cosmos*, I was greatly pleased to learn that there was a serious concern in England to construct Mr Charles Babbage's *analytical machine*. In the speech in which Count Rosse lists the steps which have already been taken towards this goal, I noticed above all the following passage:[b] 'Before I took any step, I wrote to several very eminent men of science, inquiring whether in their opinion any great scientific object would be gained, if Mr Babbage's views, as explained in Menabrea's little essay, were completely realized. The answers I received were strongly in the affirmative.'

Since I am not altogether unacquainted with this question, allow me, sir, to transmit to you a few details which are not without interest to those who concern themselves with science. I am the true author of the little piece which Count Rosse attributes to Mr Babbage under the pseudonym of Menabrea.

---

[a] This letter is a new English translation of the preceding article in this volume.

[b] Earl of Rosse, PRS, 'Address of the President', *Proceedings of the Royal Society*, Vol. 7 (1855), pp. 248–63; the quotation has been taken from the address, rather than retranslating the French.

A fair number of years ago, Mr Babbage in the course of a journey in Italy, paused for a while in Turin, where he was so good as to explain to me the main features of his *analytical machine*, which is fundamentally different from the *difference machine* already known to the public. To tell the truth, the problem which Mr Babbage set himself was so singular that my first reaction was that of doubt. But on consideration, I succeeded in convincing myself that this famous sage's idea was perfectly rational and, with the consent of the author, I took it upon myself to make known the fundamental principles on which the amazing instrument concerned is based, in an article which was published in the *Bibliothèque universelle de Genève*, no. 82, October, 1842, page 352 onwards.

Some time afterwards, there appeared an English translation of this same work entitled: *Sketch of the Analytical Engine, invented by Charles Babbage, Esq., by L. F. Menabrea of Turin, officer of the Military Engineers*, with notes by the translator. (*Extracted from the Scientific Memoirs, Vol.* III. London, *printed by Richard and John E. Taylor, Red Lion Court Fleet Street*, 1843.) The notes which accompanied the translation of my little article were most remarkable and displayed, in their author, an extraordinary wisdom. I did not know the name of this author until, to my great surprise, I learned from Mr Babbage himself that the translation and the notes were the work of Lady Ada Lovelace, the daughter of Lord Byron, a lady as distinguished by the elevation of her mind as by her beauty, and who died, a few years ago, at the most brilliant period of her life.

I thought that this work had been forgotten; but, since it has recently been the starting point for the overtures made to the English Government to construct Mr Babbage's invention, I feel that I should take the opportunity to pay homage to the memory of the daughter of a great poet and to draw the attention of scholars not to my modest work, but to the notes and commentaries to be found in the translation and which are such as to make known the aim and the power of the *analytical machine*. This invention bears no resemblance to anything hitherto conceived, and I could not make known its principles, within the confines of a simple letter; but suffice to say that through it, the series of analytical and numerical operations which are necessary for the solution of a defined problem, can be executed in the same way that the Jacquard loom, with the use of cards, can create the design of a brocade or figured cloth.

<Subscription.>

L. F. MENABREA

Member of the Académie Royale des sciences and the Sardinian Parliament

X

REPORT OF THE COMMITTEE[a] . . . APPOINTED TO CONSIDER THE ADVISABILITY AND TO ESTIMATE THE EXPENSE OF CONSTRUCTING MR BABBAGE'S ANALYTICAL MACHINE, AND OF PRINTING TABLES BY ITS MEANS

BY C. W. MERRIFIELD

From *Report of the British Association for the Advancement of Science, Dublin, August, 1878* (London, 1879), pp. 92–102

We desire in the first place to record our obligations to General Henry Babbage for the frank and liberal manner in which he has assisted the Committee, not only by placing at their disposal all the information within his reach, but by exhibiting and explaining to them, at no small loss of time and sacrifice of personal convenience, the machinery and papers left by his father, the late Mr Babbage. Without the valuable aid thus kindly rendered to them by General Babbage it would have been simply impossible for the Committee to have come to any definite conclusions, or to present any useful report.

We refer to the chapter in Mr Babbage's 'Passages from the Life of a Philosopher', and to General Menabrea's paper, translated and annotated by Lady Lovelace, in the third volume of Taylor's 'Scientific Memoirs', for a general description of the Analytical Engine.

[a] The committee consisted of 'Professor Cayley, Dr Farr, Mr J. W. L. Glaisher, Dr Pole, Professor Fuller, Professor A. B. W. Kennedy, Professor Clifford, and Mr C. W. Merrifield'.

I. *The general principles of calculating engines*

The application of arithmetic to calculating machines differs from ordinary clockwork, and from geometrical construction, in that it is essentially discontinuous. In common clockwork, if two wheels are geared together so as to have a velocity ratio of 10 to 1 (say), when the faster wheel moves through the space of one tooth, the angular space moved through by the slower wheels is one-tenth of a tooth. Now in a calculating machine, which is to work with actual figures and to print them, this is exactly what we *don't* want. We require the second wheel not to move at all until it has to make a complete step, and then we require that step to be taken all at once. The time can be very easily read from the hands of a clock, and so can the gas consumption from an ordinary counter; but a moment's reflection will show what a mess any such machinery would make of an attempt at printing.

This necessity of jumping discontinuously from one figure to another is the fundamental distinction between calculating and numbering machines on the one hand, and millwork or clockwork on the other. A parallel distinction is found in pure mathematics, between the theory of numbers on the one hand, and the doctrine of continuous variation, of which the differential calculus is the type, on the other. A calculating machine may exist in either case. The common slide-rule is, in fact, a very powerful calculating machine in which the continuous process is used, and the planimeter is another.

Geometrical construction, being essentially continuous, would be quite out of place in the calculating machine which has to print its results. Linkwork also, for the same reason, is out of place as an auxiliary in any form to the calculation. It may be of service in simplifying the construction of the machine; but it must not enter into the work as an equivalent for arithmetical computation /

The primary movement of calculating engines is the *discontinuous train*, of which one form is sketched in the accompanying diagram

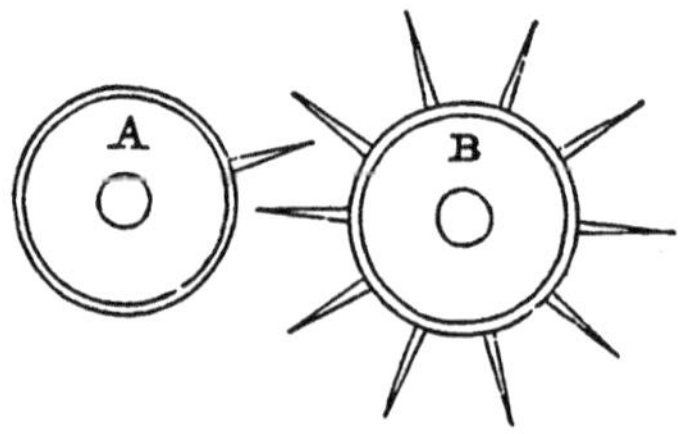

Figure 1

(Figure 1): B is the follower, an ordinary spur wheel with (say) 10 teeth; A is its driver, and this has only a single tooth. With a suitable proportion of parts, the single tooth of A only moves B one interval for a whole revolution of A; for it only gears with B by means of this single tooth. When that is not in gear, A simply slips past the teeth of B without moving the latter.

All the other machinery of calculating engines leads up to and makes use of this, or of some transformation of it, as its means of dealing with units of whatever decimal rank, instead of allowing indefinite fractions of units to appear in the result which has to be printed from.

The primary operation of calculation is counting: the secondary operation is addition, with its counterpart, subtraction. The addition and subtraction are in reality effected by means of counting, which still remains the primary operation; but the necessity for economizing labour and time forces upon us devices for performing the counting processes in a summary manner, and for allowing several of them to go on simultaneously in the calculating engine. For, if we use simple counting as our only operation, and suppose our engine set to 2312 (say), then, in order to add 3245 to it by mere repetition, we have 3245 unit operations to perform, and this is practically intolerable. If, however, we can separate the counting, so as to count on units to units only, tens to tens only, hundreds to hundreds only, and so forth, we shall only have

$$3 + 2 + 4 + 5 = 14$$

turns of the handle, as against 3245 turns. In general terms the number of operations will be measured by the sum of the digits of the number, instead of by the actual number itself. This is exactly analogous to what we should do ourselves in ordinary arithmetic in working an addition sum, if we had not learnt the addition table, but had to count on our fingers in order to add. This statement of the work is, however, incomplete. In the first place the convenience of machinery obliges us to provide 10 steps for each figure, whatever it may be, and there must be an arrangement by which the setting of the figure to be added shall cause a wheel to *gain ground* by so many steps as the number indicates, and to *mark time* without gaining ground for the other steps up to 10. Thus, in adding 7 our driver must make a complete turn or 10 steps, equivalent to 1 step of the follower; but only 7 of these steps of the driver must be effective steps, the others being skipped steps. There are various devices for this. One of

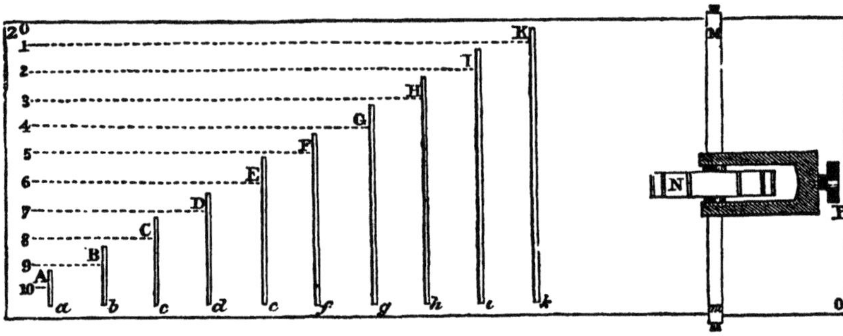

Figure 2

the simplest and most direct is that used in Thomas's / 'Arithmomètre';* another is the Reducing Bar used by Mr Babbage.† In the second place, the *carrying* has to be provided for just as in ordinary addition of numbers. Taking account of all this, it follows that by separating the counting on the whole into counting on figure by figure, / the number of separate steps is reduced from that expressed

* Let *ZO* (Figure 2) be a plate with ten ribs of different lengths, *Aa*, *Bb*, . . . *Kk* soldered on it. Let *Mm* be a square axis on which the wheel *N* is made to slide by the fork *P*. Then, supposing *N* to have teeth which can engage in the ribs *Aa*, etc., when the plate is pushed past the wheel *N*, the number of teeth through which the wheel *N*, carrying with it the shaft *Mm*, is made to rotate, depends upon the number of ribs in which it engages, and this depends upon how far along the axis *N* is made to slide by means of the fork *P*. If this fork is set opposite the line marked 3, *Mm* will turn through a space equivalent to 3 teeth. If a wheel, keyed to the shaft *Mm*, be geared to other wheels, this enables us to add any digit to any number at a single motion of the plate, by simply changing the position of *P* to suit the digit required. This is the principle used in Thomas's Arithmomètre, only that there the traversing plate is replaced by a rotating cylinder.

† Suppose *Aa, Bb*, . . . *Ff* (Figure 3) to be a series of racks passing through mortices in a plate *xz*, and meeting a series of spur-wheels mounted loose on a shaft, so that each wheel gears with one of the racks at the line *pq*, and that all the whole series can be thrown in or out of gear together. Starting with them out of gear, let the racks be drawn out through the plate *xz* as indicated. Next throw the shaft *pq* into gear, and then press a plate *PQ* against the ends of the racks, pushing them back until the plates *PQ* and *xz* meet. Then each wheel on *pq* will turn through the number of teeth corresponding to the original projection of the racks. In this way, if the wheels on *pq* stood at any given number, say 543243, we should have added 314236 to them, and they would then stand at the sum of these two numbers, namely, 857479. This, it will be observed, makes no provision for carrying. *PQ* is the reducing bar. In practice the arrangement is usually circular, the bar *PQ* revolving about an axis parallel to itself instead of sliding. If the numbers on the wheels *pq* are placed one way we get addition; if reversed, subtraction. Otherwise we may reverse by introducing an additional set of wheels between the wheels *pq* and the racks.

This is the bare principle, admitting of many transformations, and making, like the other, no provision for carrying.

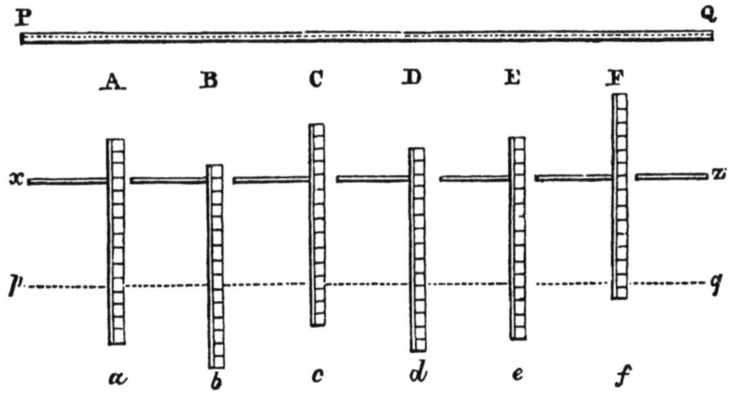

Figure 3

by the number itself to eleven times the number of its digits; that is to say, for example, the addition of the number 73592 to any other number is reduced from 73592 to 55 steps, and although of this latter some are slipped, there is no gain of time thereby, except in so far as several of the steps may be made simultaneously. The ordinary engines beat the human calculator in respect of adding all the figures simultaneously; but Mr Babbage was the first to devise a method of performing all the carrying simultaneously too.

Mechanical invention has not yet gone beyond the reduction of the distinct steps involved in the addition of a number consisting of n digits to less than $11n$; practically, from the necessity of accompanying the carrying with a *warning* step, rather more are required.

In all the calculating machines at present known, including Mr Babbage's Analytical Engine, multiplication is really effected by repeated addition. It is true that, by a multiplication of parts, more than one addition may be going on simultaneously; but it yet remains true, as a matter of mechanism, that the process is purely one of iterated addition.

By means of reversing wheels or trains, subtraction is as easily and directly performed as addition, and that without becoming in any degree a tentative process. But it is important to observe that the process can be made tentative, so as to give notice when a minuend is, or is about to become, exhausted. This is the necessary preparation for division, which is thus essentially a tentative process. That does not take it out of the power of the machine, because the machine may be, and is, so devised as to accept and act upon the notice. Nevertheless it is a step *alieni generis* from the direct process of addition, multiplication, and subtraction. It need hardly be stated that the process of obtaining a quotient consists in counting the number of subtractions employed, up to the machine giving notice of the minuend being exhausted.

Another essentially distinct train is involved in the decimal shift of the unit, in all the four elementary rules. This is most simply and most commonly effected by the sliding of an axis or frame longitudinally, after the manner of a common sliding-scale or rule, so as to bring either the figures, or the teeth which represent them, against those to which their decimal places correspond, and to no others. In multiplication and division, this means a shift for each step of the multiplication and division.

II. *Special characteristics of Mr Babbage's Analytical Engine*

1. *The mill.* The fundamental operation of Mr Babbage's Analytical Engine is simple addition. This and the other elementary rules of subtraction, multiplication, and division, and all combinations of these, are performed in what is called '*the mill*''. All the shifts which have to take place, such as changing addition into subtraction by throwing a reversing train into gear, or the shift of the decimal place, carrying and borrowing, and so forth, are effected by a system of rotating cams acting upon or / actuated by bell-cranks, tangs, and other similar devices commonly used in shifting machinery, sometimes under the name of clutches or escapements. These clutches and bell-cranks control the purely additive and carrying processes effected in the additive trains described in the note to section I, and, *being themselves suitably directed*, secure that the proper processes shall be performed upon the proper subject-matter of operation, and duly recorded, or used, as may be required.

2. *The store.* A series of columns, each containing a series of wheels, constitutes the store. This store, which may be in three or more dimensions, both receives the results of operations performed in the mill, and serves as a store for the numbers which are to be used in the mill, whether as original or as fresh subjects of operation in it. Each column in the store corresponds to a definite number, to which it is set either automatically or by hand, and the number of digits in

this number is limited by the number of wheels carried on the shaft of the column. The wheels gear into a series of racks, which can be thrown into or out of gear by means of the cards.

3. *Variable cards.* All the numbers which are the subject of operation in the mill, whether they are the result of previous operations therein, or new numbers to be operated upon for the first time, are introduced to it in the form of Jacquard* cards, such as are used in weaving. One set of wires or axes transfers the numbers on these cards to the subject of operation in the mill, exactly as similar cards direct which of the warp threads are to be pushed up, and which down, in the Jacquard loom. The mill itself punches such cards when required.

4. *Operation cards.* A different set of cards selects and prescribes the sequence of operations. These act, not upon the number wheels of the mill or store, but upon the cams and clutches which direct the gearing of these wheels and trains. Thus, in such an operation as $(ab + c)d$, we should require:

1. 4 variable cards with the numbers a, b, c, d.
2. An operation card directing the machine to multiply a and b together.
3. A record of the result, namely the product $ab = p$, as a fifth variable card.
4. An operation card directing the addition of p and c.
5. A record of the result, namely the sum $p + c = q$, as a sixth variable card.
6. An operation card directing the machine to multiply q and d together.
7. A record of the result, namely the product $qd = p_2$, either printed as a final result or punched in a seventh variable card. /

* In a letter written by Mr Babbage to Arago in December, 1839, the following explanation of the use of these cards is given. It probably conveys the idea in the fewest words possible. It is only necessary to add that their twofold employment embodies the separation of the symbols of operation from those of quantity. 'You are aware that the system of cards which Jacquard invented are the means by which we can communicate to a very ordinary loom orders to weave any pattern that may be desired. Availing myself of the same beautiful invention, I have by similar means communicated to my calculating engine orders to calculate any formula however complicated; but I have advanced one stage further, and I have communicated through the same means orders to follow certain laws in the use of those cards, and thus the calculating engine can solve any equations, eliminate between any number of variables, and perform the highest operations of analysis.'

III. *Capability of the engine*

It has already been remarked that the direct work of the engine is a combination and repetition of the processes of addition and subtraction. But in leading up to any given datum by these combinations, there is no difficulty in ascertaining tentatively when this datum is reached, or about to be reached. This is strictly a tentative process, and it appears probable that each such *tentamen* requires to be specially provided for, so as to be duly noted in the subsequent operations of the machine. There is, however, no necessary restriction to any particular process, such as division; but any direct combination of arithmetic, such as the formation of a polynomial, can be made to lead up to a given value in such a manner as to yield the solution of the corresponding equation. In any such process, however, it is evident that there can be only (to choose a simile from mechanism) one degree of freedom; otherwise the problem would yield a locus, indeterminate alike in common arithmetic, and as regards the capabilities of the machine. The possibility of several roots would be a difficulty of exactly the same character as that which presents itself in Horner's solution of equations, and the same may be said of imaginary roots differing but little from equality. These, however, are extreme cases, with which it is usually possible to deal specially as they arise, and they need not be considered as detracting materially from the value of the engine. Theoretically, the grasp of the engine appears to include the whole synthesis of arithmetic, together with one degree of freedom tentatively. Its capability thus extends to any system of operations or equations which leads to a single numerical result.

It appears to have been primarily designed with the following general object in view – to be coextensive with numerical synthesis and solution, without any special adaptation to a particular class of work, such as we see in the Difference Engine. It includes that *à majori*, and it can either calculate any single result, or tabulate any consecutive series of results just as well. But the absence of any speciality of adaptation is one of the leading features of the design.

Mr Babbage had also considered the indication of the passage through infinity as well as through zero, and also the approach to imaginary roots. For details upon these points we must refer to his 'Passages from the Life of a Philosopher'.

IV. *Present state of the design*

The only part of the Analytical Engine which has yet been put together is a small portion of 'the mill', sufficient to show the methods of addition and subtraction, and of what Mr Babbage called his 'anticipating carriage'. It is understood that General Babbage will (independently of this report) publish a full account of this method. No further mention of it will therefore be made here.

A small portion of the work is in gunmetal wheels and cranks, mounted for the most part on steel shafts. But the greater part of the wheels are in a sort of pewter hardened with zinc. This was adopted from motives of economy. They are for the most part not cast, but moulded by pressure, and the moulds of most of them are in existence.

A large number of drawings of the machinery are also in existence. It is supposed that these are complete to the extent of giving an account of / every particular movement essential to the design of the engine; but, for the most part, they are not *working drawings*, that is to say, they are not drawings suited to be sent straight to the pattern or fitting shop, to be rendered in metal. There are also drawings for the erection of the engine, and there appears to be a complete set of descriptive notes of it in Mr Babbage's 'mechanical notation'. There remains, however, a great deal to be done in the way of calculating quantities and proportions, and in the preparation of working drawings, before any work could actually be set in hand, even if the design be really complete. There is some doubt on this point as the matter stands, and it certainly would be unsafe to rely upon the design being really complete, until the working drawings had been got out. Mechanical engineers are well aware that no complex design can be trusted without this test, at least.

It was Mr Babbage's rule, in designing mechanism, in the first place to work to his object, in utter disregard of any questions of complexity. This is a good rule in all devising of methods, whether analytical, mechanical, or administrative. But it leaves in doubt, until the design finally leaves the inventor's hands in a finished state, whether it really represents what is meant to be rendered in metal, or whether it is simply a provisional solution, to be afterwards simplified.

V. *Probable cost*

It has not been possible for us to form any exact conclusion as to the cost. Nevertheless there are some data in existence which appear to fix a lower limit to the cost. Mr Babbage, in his published papers, talks of having 1000 columns of wheels, each containing 50 distinct wheels; this apparently refers to his *store*. Besides the many thousand moulded pewter wheels for these, and the axes on which they are mounted, there is the *mill*, also consisting of a series of columns of wheels and of a vast machinery of cams, clutches, and cranks for their control and connection, so as to bring them within the directing power of the Jacquard systems of variable cards and operation cards. Without attempting any exact estimate, we may say that it would surprise us very much if it were found possible to obtain tenders for less than £10,000, while it would pretty certainly cost a considerable sum to put the design in a fit state for obtaining tenders. On the other hand, it would not surprise us if the cost were to reach three or four times the amount above suggested.

It is understood that towards the close of his life Mr Babbage had contemplated carrying out the manufacture of the engine on a smaller scale, confining himself to 25 figures instead of 50, and to 200 columns instead of 1000 or more. This would of course reduce the amount of the metal work proportionately, but we do not think that it would materially reduce the charge which we anticipate for bringing the design into working order.

VI. *Strength and durability*

The questions of strength and durability had by no means escaped Mr Babbage's attention, and a great deal of his detail bears marks of having been designed with especial reference to these two points. That was essential in a large and complex engine with some thousands of wheels, all requiring at some time or other, although not simultaneously, to be driven by the means of one shaft. This necessarily throws a great deal of pressure, and also a great deal of wear and tear, on the main driving shaft / and the gear immediately connected with it. We have no means of knowing, in the present state of the design, to what extent Mr Babbage had succeeded in reducing this, or whether he had always been successful in arranging his cams and cranks so as to secure the best working angles, and to avoid their being jammed at dead points or otherwise. Giving him full credit for being quite aware of the importance of this, we cannot but doubt whether the design was ever in a sufficiently forward state to enable him, or anyone else, to speak with certainty on this point. Several of the existing calculating machines show signs of weakness in the driving pinions.

One of the movements apparently necessary to the tentative processes of the engine is, when the spur-wheels on a given shaft have been brought into certain definite positions depending on previous operations, to bring up a sharp straight edge against them in a plane passing through the axis of the shaft. This pushes some to right and others to left, according to the position of the crown of the tooth relatively to the straight edge. This operation is necessary to secure that the clearance of the different parts of the machinery, whether originally provided in order to allow it to work smoothly, or whether afterwards increased by working, shall not introduce a numerical error into the result. The principle of this operation is used generally throughout the Analytical Engine. Its consequent effect, both in respect of the work which it throws upon the main driving gear, and of the wear of the parts which it pushes, forms an important element in considering the durability of the machine. This bar also serves the purpose of locking part of the machine when required.

On the other hand, it is to be remarked, that the use of springs has been wholly discarded by Mr Babbage, as directors of motion, although he occasionally uses them for return motions.

VII. *Probable utilization of the Analytical Engine*

It has been already remarked that one of the main features of the engine is, that its function is coextensive with numerical synthesis and solution, and that there is an absence of any special adaptation. In thus widening the sphere of its capability, it is made to diverge from the general tendency of mechanical design, which is towards the selection and particularization of the work to be performed, and the restriction of the machinery to one particular cycle of operation, usually within close numerical limits, as well as limited in kind. Nevertheless, modern engineering practice finds ample room for 'universal' drills, shaping tools, and other machines having very general adjustments and applications. But it remains practically true that each step of freedom of adjustment is also a step in diminution of special aptitude.

Whilst the Analytical Engine is capable of turning out a single result, as the combination of a complex series of numbers and operations performed upon them, it can also yield a series of such results in a consecutive form, and thus give tabulated results. Only it is not restricted, as is the Difference Engine, to the special method of tabulation by finite differences, nor is tabulation its primary function or intention. If its actual capabilities are found to realize the intentions of its inventor, it will tabulate all functions which are within the reach of numerical synthesis, and those direct inversions of it which are known under the name of solutions. It deals, however, with number, and not with analytical form. /

Theoretically it might supersede the Difference Engine, *à majori*; but for reasons already stated, the specialization of the Difference Engine would probably give it an advantage over the more powerful engine, when the work was specially suited to finite differences.

There would remain much work, tabular and other, for which differences are not very directly suited. Amongst these may be mentioned the determination of heavy series of constants and of definite functions of them, such as Bernoulli's numbers, Σx^{-n}, coefficients of various expansions of functions, and inversions of known expansions, solutions of simultaneous equations with large numerical coefficients and many variables, including, as a particular, but important case, the practical correction of observations by the method of least squares. If all sorts of heavy work of this kind could be easily and quickly, as well as certainly, done, by merely selecting or punching a few Jacquard cards and turning a handle, not only much saving of labour would result, but much which is now out of human possibility would be brought within easy reach.

If intelligently directed and saved from wasteful use, such a machine might mark an era in the history of computation, as decided as the introduction of logarithms in the seventeenth century did in trigonometrical and astronomical arithmetic. Care might be required to guard against misuse, especially against the imposition of Sisyphean tasks upon it by influential sciolists. This, however, is no more than has happened in the history of logarithms. Much work has been done with them which could more easily have been done without them, and the old reproach is probably true, that more work has been spent upon making tables than has been saved by their use. Yet, on the whole, there can be no reasonable doubt that the first calculation of logarithmic tables was an expenditure of capital which has repaid itself over and over again. So probably would the Analytical Engine, whatever its cost, if we could be assured of its success.

VIII. *Possible modification of the engine*

Without prejudging the general question referred to us as to the advisability of completing Mr Babbage's engine in the exact shape in which it exists in the machinery and designs left by its inventor, it is open to consideration whether some modification of it, to the sacrifice of some portion of its generality, would not reduce the cost, and simplify the machinery, so as to bring it within the range of both commercial and mechanical certainty. The 'mill', for example, is an exceedingly good mechanical arrangement for the operations of addition and subtraction, and with a slight modification, with or without store columns, for multiplication. We have already called attention to the imperfection of the existing machines, which show weakness and occasional uncertainty. It is at least worth consideration whether a portion of the Analytical Engine might not thus be advantageously specialized, so as to furnish a better multiplying machine than we at present possess. This, we have reason to believe, is a great desideratum both in public and private offices, as well as in aid of mathematical calculators.

Another important desideratum to which the machine might be adapted, without the introduction of any tentative processes (out of which the complications of the machinery chiefly arise) is the solution of simultaneous equations containing many variables. This would include / a large part of the calculations involved in the practical application of the method of least squares. The solution of such equations can always be expressed as the quotient of two determinants, and the obtaining this quotient is a final operation, which may be left to the operator to perform by ordinary arithmetic, or which may be the subject of a separate piece of machinery, so that the more direct work of forming the determinant, which is a mere combination of the three direct operations of addition, subtraction, and multiplication, may be entirely freed from the tentative process of division, which would thus be prevented from complicating the direct machinery. In the absence of a special engine for the purpose, the solution of large sets of simultaneous equations is a most laborious task, and a very expensive process indeed, when it has to be paid for, in the cases in which the result is imperatively needed. An engine that would do this work at moderate cost would place a new and most valuable computing power at the disposal of analysts and physicists.

Other special modifications of the engine might also find a fair field for reproductive employment. We do not think it necessary to go into these questions at any great length, because they involve a departure, in the way of restriction and specialization, from Mr Babbage's idea, of which generality was the leading feature. Nevertheless, we think that we should be guilty of an omission, if we were to fail to suggest them for consideration.

IX. *General conclusions, and recommendation*

1. We are of opinion that the labours of Mr Babbage, firstly on his Difference Engine, and secondly on his Analytical Engine, are a marvel of mechanical ingenuity and resource.
2. We entertain no doubt as to the utility of such an engine as was in his contemplation when he undertook the invention of his Analytical Engine, supposing it to be successfully constructed and maintained in efficiency.
3. We do not consider that the possibilities of its misuse are any serious drawback to its use or value.
4. Apart from the question of its saving labour in operations now possible, we think the existence of such an instrument would place within reach much which, if not actually impossible, has been too close to the limits of human skill and endurance to be practically available.
5. We have come to the conclusion that in the present state of the design of the engine it is not possible for us to form any reasonable estimate of its cost, or of its strength and durability.
6. We are also of opinion that, in the present state of the design, it is not more than a theoretical possibility; that is to say, we do not consider it a certainty that it could be constructed and put together so as to run smoothly and correctly, and to do the work expected of it.
7. We think that there remains much detail to be worked out, and possibly some further invention needed, before the design can be brought into a state in which it would be possible to judge whether it would really so work.

8. We think that a further cost would have to be incurred in order to bring the design to this stage, and that it is just possible that a mechanical failure might cause this expenditure to be lost.

9. Whilst we are unable to frame any exact estimates, we have reason / to think that the cost of the engine, after the drawings are completed, would be expressed in tens of thousands of pounds at least.

10. We think there is even less possibility of forming an opinion as to its strength and durability than as to its feasibility or cost.

11. Having regard to all these considerations, we have come, not without reluctance, to the conclusion, that we cannot advise the British Association to take any steps, either by way of recommendation or otherwise, to procure the construction of Mr Babbage's Analytical Engine and the printing tables by its means.

12. We think it, however, a question for further consideration whether some specialized modification of the engine might not be worth construction, to serve as a simple multiplying machine, and another modification of it arranged for the calculation of determinants, so as to serve for the solution of simultaneous equations. This, however, inasmuch as it involves a departure from the general idea of the inventor, we regard as lying outside the terms of reference, and therefore perhaps rather for the consideration of Mr Babbage's representatives than ours. We accordingly confine ourselves to the mere mention of it by way of suggestion. /

XI

ON THE MECHANICAL ARRANGEMENTS OF THE ANALYTICAL ENGINE[a]

BY H. P. BABBAGE

From *Babbage's calculating engines* (London, 1889)

Ten years have elapsed since a committee of the British Association reported[b] upon the Analytical Engine of my father, Charles Babbage, and I desire now, whilst offering a few remarks upon that report, to endeavour to convey some idea of the mechanical arrangements of the engine to those who may be interested in it.

2. I am well assured that a time will come when such an engine will be completed and be a powerful means of enlarging not only pure mathematical science, but other branches of knowledge, and I wish, as far as in me lies, to hasten that time, and to help towards the general appreciation of the labours of my father, so little known or understood by the multitude even of the educated.

3. He considered the paper by Menabrea, translated with notes by Lady Lovelace, published in volume 3 of Taylor's 'Scientific Memoirs',

[a] This paper was read by H. P. Babbage to the meeting of the British Association, Bath, 12 September, 1888. An abstract of the paper appeared in the *Report of the British Association for the Advancement of Science* (1888), pp. 616–7. The full text of the paper was printed by H. P. Babbage in *Babbage's calculating engines* (London, 1889), pp. 331–7. Footnotes in the original have been interpolated to conform to the pagination of the *Works of Babbage*. In some footnotes, which originally cited diagrams and notations not reprinted in the *Works of Babbage*, the reader is referred to *Babbage's calculating engines*; the omission of these items will not affect the reader's understanding of the paper.

[b] C. W. Merrifield, 'Report of the committee . . . appointed to consider the advisability and to estimate the expense of constructing Mr Babbage's analytical machine, and of printing tables by its means' (1879), reprinted in this volume.

as quite disposing of the mathematical aspect of the invention. My business now is not with that.

4. The idea of the Analytical Engine arose thus: When the fragment of the Difference Engine, now in the South Kensington Museum, was put together early in 1833, it was found that, as had been before anticipated, it possessed powers beyond those for which it was intended, and some of them could be and were demonstrated on that fragment.*

5. It was evident that by interposing a few connecting-wheels, the column of result can be made to influence the last difference, or other part of the machine in several ways. Following out this train of thought, he first proposed to arrange the axes of the Difference Engine circularly, so that the result column should be near that of the last difference, and thus easily within reach of it.† He called this arrangement 'the engine eating its own tail'.‡ But this soon led to the idea of controlling the machine by entirely independent means, and making it perform not only addition, but all the processes of arithmetic at will in any order and as many times as might be required. Work on the Difference Engine was stopped on 10 April, 1833, and the first drawing of the Analytical Engine is dated in September, 1834.

6. The object may shortly be given thus: It is a machine to calculate the numerical value or values of any formula or function of which the mathematician can indicate the *method* of solution. It is to perform the ordinary rules of arithmetic in any order as previously settled by the mathematician, and any number of times and on any quantities.

7. It is to be absolutely automatic, the slave of the mathematician, carrying out his orders and relieving him from the drudgery of computing.

8. The Analytical Engine is of course to print the results, or any intermediate result arrived at. He regarded this as an indispensable requisite, without which a calculating machine might indeed be useful for some purposes, but not for those of any scientific value. The perpetual risk of error in copying and transferring lines of figures is most troublesome, and tends to make results, themselves perfectly accurate, unreliable in use.

9. It is at once seen that the necessity of the engine being

* See <*Ninth Bridgewater treatise*, parts of chapters II, VIII, and XIII, and chapter XV>.

† See <*Passages*, *Works of Babbage*, Vol. 11, p. 85>.

‡ <Refer to *Babbage's calculating engines*.>

automatic imposes a gigantic task on the inventor. The first means employed to meet it is the use of cards to govern the engine. These are very similar to those in use in the Jacquard loom, to which we owe the figured patterns in the beautiful fabrics we see everywhere in common use.

10. One set of cards could be used to communicate the 'given numbers', or constants of a problem to the machine. I shall call these throughout this paper 'Number Cards'.*

11. Another set of cards would be used to direct to which particular place or column in the engine these numbers, or any intermediate numbers arising in the course of the calculation, are to be conveyed / or transferred; these cards I will call 'Directive Cards'. There would also be other 'Directive Cards' for general purposes of control when necessary.

12. A third sort, called 'Operation Cards',† would direct the actual operations to be performed, these would put the engine mechanically into a condition to perform the particular operation required – addition, subtraction, etc., etc.

13. I was once asked, 'How do you set the question? Do you write it on paper and put it into the machine?' Well, given the problem, the mathematician must first of all settle the operations and the particular quantities each is to be performed on and the time for each operation.

14. Then the superintendent of the engine must make a 'Number Card' for each 'given number', and settle the particular column in the machine on to which each 'given number' is to be first received, and assign columns for every intermediate result expected to arise in the course of the calculation.

15. He will then prepare 'Directive Cards' accordingly, and these, together with the necessary 'Operation Cards' being placed in the engine, the question will have been set; not exactly, as my friend suggested, written on paper, but in cardboard, and motion being supplied the engine will give the answer.‡

16. Now this appears a long process for what may be a simple question, but it is to be noted that the engine is designed for analytical purposes, and it would be like using the steam hammer to crush the nut, to use the Analytical Engine to solve common sums in

* See diagram on general plan, No. 25 of 6 August, 1840 <at end of this volume>.

† See <*Passages*, p. 96> and general plan of 6 August, 1840.

‡ See <*Passages*, p. 89>.

arithmetic; or, adopting the language of Liebnitz:* 'It is not made for those who sell vegetables or little fishes, but for observatories, or the private rooms of calculators, or for others who can easily bear the expense, and need a good deal of calculation.' Moreover, except the 'Number Cards', all the cards, once made for any given problem, can be used for the same problem, with any other 'given numbers', and it would not be necessary to prepare them a second time – they could be carefully kept for future use. Each formula would require its own set of cards, and by degrees the engine would have a library of its own.†

17. Thus the values for any number of life insurance policies might be calculated one after the other by merely supplying fresh cards for the age, amount, rate of interest, etc., for each individual case.

18. The separation of 'Operation Cards' from the numbers to be operated on is complete. The powers of the engine are the most extended, but each set of cards makes it special for the solution of one particular problem; each individual case of which, again, requires its own 'Number Cards'.

19. Taking the formula used by Mr Merrifield $(ab + c)d$ as an illustration,‡ the full detail of the cards of all sorts required, and the order in which they would come into play is this:

Directive Card	*Operation Card*	
1st	..	Places a on column 1 of store
2nd	..	Places b on column 2 of store
3rd	..	Places c on column 3 of store
4th	..	Places d on column 4 of store
5th	..	Brings a from store to mill
6th	..	Brings b from store to mill
..	1	Multiplies a and $b = p$
7th	..	Takes p to column 5 of store where it is kept for use and record
8th	..	Brings p into mill
9th	..	Brings c into mill
..	2	Adds p and $c = q$
10th	..	Takes q to column 6 of store
11th	..	Brings d into mill
12th	..	Brings q into mill
..	3	Multiplies $d \times q = p_2$
13th	..	Takes p_2 to column 7 of store
14th	..	Takes p_2 to printing or stereomoulding apparatus

* See <*Works of Babbage*, Vol. 2, p. 180>.
† See <*Passages*, p. 90>.
‡ See <this volume, p. 181>.

The four cards for the 'given numbers' *a, b, c* and *d*, strung together are placed by hand on the roller, these numbers have to be placed on the columns assigned to them in a part of the machine called 'The Store', where every quantity is first received and kept ready for use as wanted.

20. We have thus besides the 'Given Number' cards, three 'Operation Cards' used, and fourteen 'Directive Cards'; each set of cards would be strung together and placed on a roller or prism / of its own; this roller would be suspended and be moved to and fro. Each backward motion would cause the prism to move one face, bringing the next card into play, just as on the loom.

21. It is obvious that the rollers must be made to work in harmony, and for this purpose the levers which make the rollers turn would themselves be controlled by suitable means, or by general 'Directive Cards', and the beats of the suspended rollers be stopped in the proper intervals.

22. This brings me to the second great distinguishing feature of the engine, the principle of 'Chain'. This enables us to deal mechanically with any single combination which may occur out of many possible, and thus to be ready for any or every contingency which may arise.

23. Supposing that it is desired to provide for a certain possible combination such as, for example, the concurrence of ten different events, it could be effected mechanically thus* – each event would be represented by an arm turning on its axis, and having at its end a block held loosely and capable of vertical motion independently of the arm which carries it. Now suppose each of these arms to be brought, on the occurrence of the event it represents, into a position so that the blocks should all be in one vertical line, then if the block in the lowest arm was raised by a lever, it would raise all the nine blocks together, and the top one could be made to ring a bell or communicate motion, etc., etc.; but if any one of the ten events had not happened, its block would be out of the 'Chain', and the lowest block would be raised in vain and the bell remain silent.

24. This is the simplest form of 'Chain', there may be many modifications of it to suit various purposes.

25. In its largest extent it will appear in the anticipating carriage† further on, but in its simplest form it appears here as a means of

* See Plate 12 of 4 December, 1836 <this volume p. 251>.

‡ See <*Passages*, p. 87>.

producing intermittent motion at uncertain intervals, which may or may not be previously known either to the mathematician or even to the superintendent of the engine who had prepared the cards.

26. In our illustration the first Operation Card happens to be multiplication, and the time this would occupy necessarily depends upon the number of digits in the two quantities multiplied. Now when multiplication is directed, the 'Chain' to every part of the machine not wanted will be broken, and all motion thus stopped till the multiplication is completed, and whenever the last step of that is reached, the 'Chain' will be restored.

27. The chain for this purpose may be made this way, a link is cut and the two ends are brought side by side, overlapping each other; a part of each link is then cut away exactly the same in both pieces. Now if the piece in communication with the driving power is moved, the other is not, and remains stationary; but if a block is let fall into the space cut away so as to fill the gaps in both parts of the severed link, the chain will be complete and link – block – link will pass on the motion. The block is hung in the end of an arm in which it can slide, and all that is required is to move the arm sideways whilst the link pieces are at rest in the proper position. This is shown in the diagram on the facing page.

28. In a machine such as the Analytical Engine, consisting of so many distinct trains of motion of which only a few would be in action at a time, it may be easily conceived how useful this application of the principle of chain is. It helps to realize, too, another important principle largely adopted, viz., to break up every train of motion as far as possible into short courses, the last step of each furnishing a mere guide for a fresh start in the mechanism from the driving power. Of course the link could be broken and restored also by any of the 'Directive Cards' mentioned above. The interposition of such links would also be used to save the cards from the wear and tear unavoidable if they were used for the actual transmission of any force.

29. I hope that I have now given some idea of the methods used for the general control of the engine, and shall pass on to illustrate, as well as I am able, some of the details of the machine. It is not to be supposed that I have mastered all these myself; nor, had I done so, would it be possible for me to make them intelligible in the course of a paper, but some idea of them I hope to convey.

30. The machine consists of many parts. I have found it easier

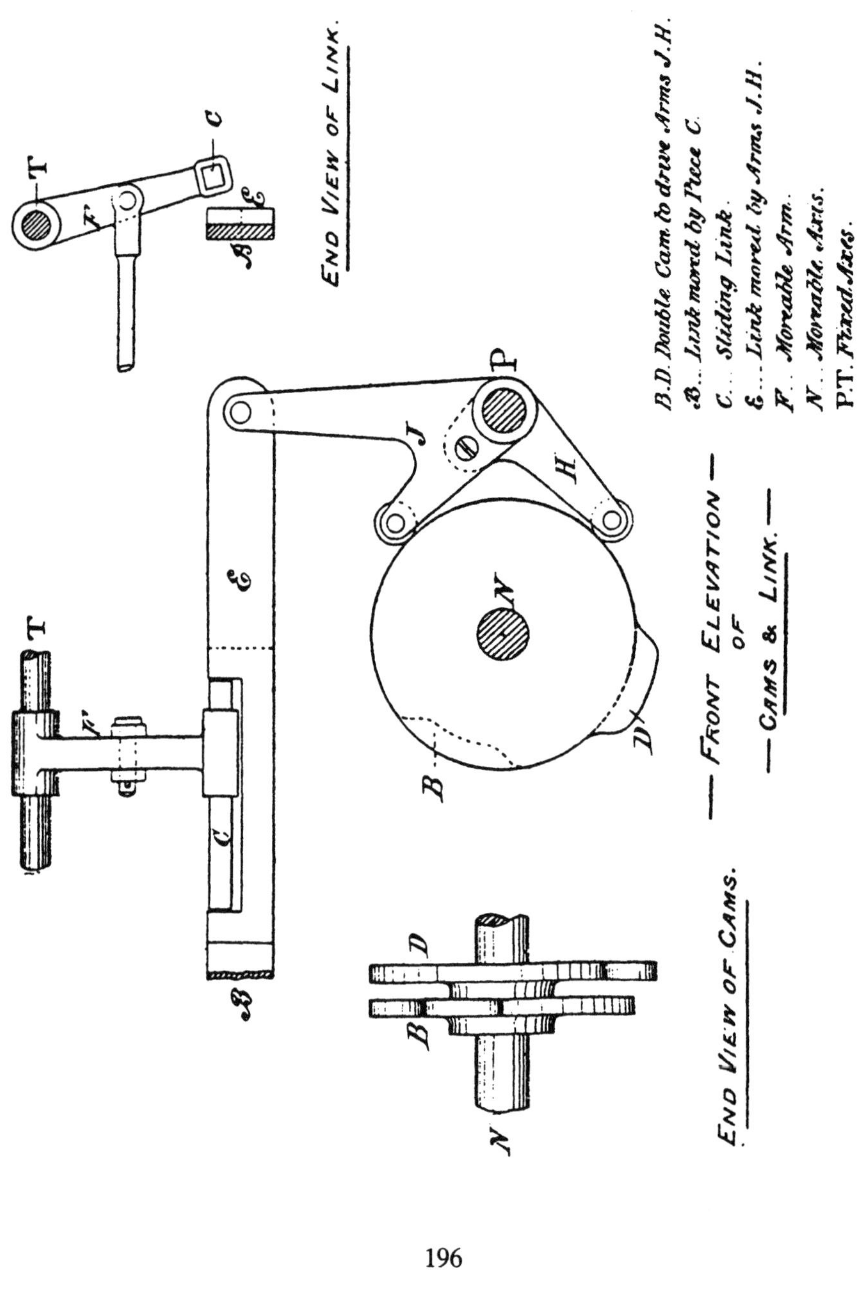
END VIEW OF LINK.
FRONT ELEVATION OF CAMS & LINK.
END VIEW OF CAMS.
B.D. Double Cam to drive Arms J.H.
B. Link moved by Piece C.
C. Sliding Link.
E. Link moved by Arms J.H.
F. Moveable Arm.
N. Moveable Axis.
P.T. Fixed Axes.

myself to regard these parts as so many separate machines, driven by the same motive power and starting and stopping each other in every possible combination, but otherwise acting independently, though with a settled harmony towards a desired result. When that idea has been reached, it seems easy to imagine them all brought close together, grouped in the positions most convenient.

31. Many general plans have been drawn. Plan No. 25, dated 6 August, 1840, was lithographed, others followed. The fact is, that what suits one part best does not suit another, and the number of possible variations of the parts renders it difficult to settle which combination of the whole may be the best for the general plan. /

32. A part of the machine called 'The Store' has already been mentioned. It would consist of a number of vertical columns to receive the 'given numbers' and those arising in the course of the calculations from them.

33. In the example already mentioned seven columns would be used; perhaps for a first machine twenty would do, with twenty-five wheels in each; each wheel might have a disc with the digits 0 to 9 engraved on its circumference, but this is not absolutely necessary. In fact, the whole machine might be constructed without a figure anywhere except for the printing.

34. The 'Directive Cards' would put the column selected to receive a 'given number' into gear with a set of racks through which the 'given number' (expressed on its card) would be conveyed to the store column, each wheel being moved as many teeth as in the corresponding place of the number.

35. One revolution of the main shaft would be sufficient to put a 'given number' on a column of the store, or to transfer it from the store to some other part of the engine.

36. The action of the 'Directive Card' in this case is to raise the selected column so that its wheels should be level with a corresponding set of racks, and thus brought into gear with them. Each column of the store would require its own 'Directive Card'.

37. At the top of each column of the store would be a wheel on which the algebraic symbol of the quantity could be written,* and on the top of those columns assigned to receive the intermediate results this would also be done; though at first these columns would be at zero.

* See <*Passages*, p. 89>.

38. There would also be a wheel for the sign + or −. Drawings and notations for such sign wheels exist,* and how far the operations of algebra may be exhibited has been discussed.

39. The mill is the part of the machine where the quantities are operated on. Two numbers being transferred from the store to two columns in the mill the two are put into gear together through racks, and the reduction of one column to zero turns the other the exact equivalent, and thus adds it to the other. Supposing there to be with each wheel a disc with the digits 0 to 9 engraved on its edge, and a screen in front of the column with a hole or window before each disc allowing one figure only to be seen at a time; during the process of addition the digits will pass before the window just as in counting till the sum is reached: thus if 5 is added to 7, the 7 will disappear and 8, 9, 0, 1, pass before the window in succession till 2 appear. At the moment when 9 passes to 0, a lever will be moved, thus recording the necessity of a carriage to the figure above; the carriage is made subsequently, and for the Analytical Engine a method of performing the carriages all simultaneously† was invented by my father which he called 'Anticipating' carriage.

40. When two numbers are added, carriages may occur in any or every place except the last; where the wheels pass as just described from 9 to 0, a carriage arises directly; in those places where the figure disc comes to rest at 9, no carriage occurs, but one of two things may happen – if there is no carriage to arrive from the next right hand place, the 9 will not be changed; but should one be due, not only the 9 must be pushed on to 0, but a carriage must be passed on to the next place on the left hand, and if there happen to be a succession of 9's, they will all have to be pushed on. Working with twenty-five places of figures there can be no carriage ever required in the last place, but there may be in any one, any two, any three, etc., up to twenty-four places, where a carriage may arise directly and the same up to twenty-three places where the presence of 9's may indirectly cause it.

41. Now immense as is the number of these two sets of combinations, they can be successfully dealt with mechanically by the principle of 'Chain'‡ and indeed every single combination as it arises is presented to the eye. There is a series of blocks two for each place, the lower block is made to serve for the two events. The upper one

* See Plate 5 <this volume p. 244 and refer to *Babbage's calculating engines*>.

† See <*Passages*, p. 87>.

‡ See Plate 12 <this volume p. 251, and refer to *Babbage's calculating engines*>.

has a projecting arm, which when moved circularly engages a toothed wheel and moves it on one tooth affecting the figure disc similarly. It rests on the lower block which moves it up and down with itself always. After the addition is completed, should a carriage have become due and the warning lever have been pushed aside, it is made (by motion from the main shaft) to actuate the lower block and throw it into 'Chain' when this is raised (again by motion from the main shaft) it raises the upper block which thus effects ordinary carriage; but supposing no carriage to have become due, there will be at the window either a 9 or some less number. The latter case may be dismissed at once; as a carriage arriving to it will cause no carriage to be passed on. Should there be a 9 at the window the warning lever cannot have been pushed aside, but in every place where there is a 9 another lever, again by motion from the main shaft comes into play and pushes the lower block into 'Chain'; not, of course, into 'Chain' for ordinary carriage, but into another position for chain for 9's, so that should there come a carriage from below, the chain for 9's will be raised as far as it extends / and effect the carriages necessary, be it for a single place, or for several, or for many. Should there be no carriage from below to disturb it the chain remains passive; it has been made ready for a possible event which has not occurred. A certain time is required for the preparation of 'Chain', but that done all the carriages are then effected simultaneously. A piece of mechanism for anticipating carriage on this plan to twenty-nine figures exists and works perfectly.

42. When a large number of places of figures is being dealt with, the saving of time is very considerable, especially when it is remembered that multiplication is usually done by successive additions.

43. Another plan for carriage has also been contrived and drawn.* It is obvious that there is no necessity, when there are many successive additions, to make the carriages immediately follow each addition. The additions may be made one after the other, and the carriages having been warned or even actually made on a separate wheel in each place as they arise, can all be made in one lot afterwards; more machinery is required, but the saving of time is very considerable. This plan has been called 'Hoarding Carriage', and thoroughly worked out.

44. It is interesting to note that hoarding carriage is seen in the

* <Refer to *Babbage's calculating engines*.>

little machine of Sir Samuel Moreland invented in 1666, and probably existed in that of Pascal still earlier.

45. Now it may happen in addition that two or more numbers being added together, there may not be room at the top of the column for the left-hand figure of the result. This would usually happen from an oversight in preparing or arranging the cards when space should be left; but it might so happen* that the calculation led, as mathematical problems sometimes do, through infinity. In either case a bell would be rung and the engine stopped; or if the contingency had been anticipated by the mathematician, fresh directive cards previously prepared might be brought into play.

46. One addition would be executed in each turn of the main axis, the intermittent motions required being produced by cams on the main axis. These would be flat discs with projecting parts on them acting on arms with friction rollers at the end. Each cam would be double, i.e. have two discs; the projections on the one coresponding with depressions on the other. Such cams are easily made and fixed and adjusted, about six or seven are sufficient for addition. The illustration shows such a double cam, together with a 'Chain' for throwing into and out of gear, as explained on page 195. This may or may not be wanted.

47. Subtraction is performed by the interposition of an additional pinion, which turns the figure discs the reverse way; the figures decreasing in succession as they pass the window, and the carriage arises whenever the 0 passes and a 9 appears. The same arrangement is applied to the 0's in subtraction as to the 9's in addition, and the same principle of 'Chain'; indeed, the same actual mechanism serves for both, the change being made by the movement of a single lever. See page 196.

48. In subtraction, when a larger is subtracted from a smaller number, there will be a warning made in the highest place for a carriage to a place above which does not exist, zero has been passed; the warning lever will ring a bell and stop the engine, unless the contingency has been anticipated and provided for by the mathematician.†

49. Several ways have been worked out‡ and drawn for multiplication. A skilled computer dealing with many places of figures having to multiply would make by successive additions a table of the first ten

* See <*Passages*, p. 100>.

† See <*Passages*, p. 100>.

‡ <Refer to *Babbage's calculating engines*.>

multiples of the multiplicand; if he has done this correctly the tenth multiple will be the same as the first, only all moved one place to the left and with a 0 in the units place. Using this table he picks out in succession the multiples required and puts them in the proper places, then adding all up gets his result, dispensing altogether with the multiplication table and doing nothing beyond addition. For the machine this way has been worked out and many drawings and notations exist for it.

50. Another way by the use of barrels* has also been drawn. The way by succession additions and stepping is perhaps the simplest.

51. When two numbers each of any number of places from one up to twenty or thirty have to be multiplied, it becomes necessary, in order to save time, to ascertain which has the fewest significant digits; special apparatus has been designed for this, called 'Digit Counting Apparatus'.† The smaller of the two is made the multiplier. Both are brought into the mill and put on the proper columns. As the successive additions are made, the figure wheels of the multiplier are successively reduced to zero; when this happens for any one figure of the multiplier a cam on its wheel pushes out a lever which breaks the link or 'chain' for addition, and completes that for stepping; so that the next revolution of the main axis causes stepping instead of addition, and that being done the links are changed back and the successive additions go on.

52. By this process the multiplier column is all reduced to zero; but if need be there can be / another column alongside to which it may step by step be worked on – anyhow, it stands on record in its own column in the store.

53. Multiplication would ordinarily be performed from the highest place downwards, and from the decimal point onwards; so that the result would be complete to the lowest place of decimals; but, of course, would contain the accumulated error due to cutting off the figures beyond in each quantity operated on. As, however, there would be a counting apparatus recording the successive additions, which at the end of the multiplication, would give the sum of all the digits of the multiplier, the maximum possible error would be known, and if thought advisable, a correction ordered to be made for it; for instance, the machine might be directed to halve the total of the

* <Refer to *Babbage's calculating engines*.>
† <Refer to *Babbage's calculating engines*.>

digits, and add it to the last figures of the result wherever cut off.

54. Division is a more troublesome operation by far than multiplication. Mr Merrifield in his report has called this 'essentially a tentative process', and so it is as regards the computer with the pen; but as regards the Analytical Engine, I do not assent to it.

55. As, however, he considered the striking part of a clock a tentative process, while I do not, the difference might be one of definition of a word, and I only notice it as maintaining that the processes of the engine are only so far tentative as the guiding spirit of the mathematician leads him to make them.

56. Division by table and also by barrels has been thoroughly drawn and worked out – the process by successive subtraction also. The divisor and dividend being brought into the mill, the successive subtractions proceed and their number is recorded; when the correct figure has been reached, the subtraction is thrown out by 'Chain' and stepping caused.

57. There is no trial and error whatever. Some additional machinery is required for this and more time is occupied, but the result is certain; the engine stops of necessity at the right figure of the quotient, and gives the order for stepping and so on to the end.

58. For the extraction of the square root a barrel would be used, but the ordinary process might be followed step by step without a barrel.

59. Counting machines of sorts would perform important functions in the general directive. Some would be mere records of the progress of the different steps going on, but others, of which the multiplier column may serve as a sort of example, would themselves act at appointed times as might be previously arranged.

60. This principle of 'Chain' is used also to govern the engine in those cases where the mathematician himself is not able to say beforehand what may happen, and what course is to be pursued, but has to let it depend on the intermediate result of the calculation arrived at. He may wish to shape it in different ways according as one or several events may occur, and 'Chain' gives him the power to do it mechanically. By this contrivance machines to play simple games of skill such as 'tit tat too' have been designed.

61. Take a simpler case where the mathematician desired to deal with the largest of two numbers arrived at, not knowing beforehand which it might be.* The numbers would have been placed on the two

* See <*Passages*, p. 100>.

columns of the store previously assigned for their reception, and cards would have been arranged to direct the two numbers to be subtracted each from the other. In the one case there would be a remainder, but in the other the carriage warning lever would have been moved in the highest place, which would be made to bring an alternative set of cards, previously prepared, into play.

62. I have not been quite able to accept Mr Merrifield's opinion as to the 'capability of the engine'. He says: 'Its capability thus extends to any system of operations or equations which leads to a single numerical result.' Now it could furnish not only one root, but every root of an equation, if there were more than one, capable of arithmetical expression, and there are many such equations. It could follow the processes of the mathematician be they tentative or direct,* wherever he could show the way to any number of numerical results. It is only a question of cards and time. Fabrics have been woven requiring several thousand cards. I possess one† made by the aid of over twenty thousand cards, and there is no reason why an equal number of cards should not be used if necessary, in an Analytical Engine for the purposes of the mathematician.

63. There exist over two hundred drawings, in full detail, to scale, of the engine and its parts. These were beautifully executed by a highly skilled draughtsman and were very costly.‡

64. There are over four hundred notations of different parts.§ These are, in my father's system of mechanical notation, an outline of which I had the honour to submit to this association at Glasgow in 1855.|| Not many years ago I was looking over one of my own drawings with a very intelligent mechanical engineer in Clerkenwell. I wished to get motion for some particular purpose. He suggested to get it from an axis which he pointed to on the drawing. I answered, 'No, that will not / do; I see by the drawing that it is a "dead centre".' He replied, 'You have some means of knowing which I have not.' I certainly had, for I used this mechanical notation. The system in whole or part should be taught in our art and technical schools.

* See <*Passages*, p. 98>.
† A portrait of Jacquard woven in silk, see <*Passages*, p. 127>.
‡ <Refer to *Babbage's calculating engines*.>
§ <Refer to *Babbage's calculating engines*.>
|| <Refer to *Babbage's calculating engines*.>

65. In addition to the above things there will shortly be available, as I have said before, the reprint of the various papers published relating to these machines, and a full list of the drawings and notations will be included.

66. I believe that the present state of the design would admit of the engine being executed in metal; nor do I think that, as suggested by Mr Merrifield, quantities and proportions would have to be calculated. Of course working drawings would have to be made – it would not be wise to commence any work without such drawings – but they would mostly be simply copies from the originals with such details as workmen want added. It would also be wise to make models of particular parts. The shapes of the cams, for instance, might be tried in wood which would afterwards suit as patterns for castings.

67. Mr Merrifield doubts 'whether the drawings really represent what is meant to be rendered in metal, or whether it is simply a provisional solution to be afterwards simplified'. I have no doubt that the drawings do represent what at the time was intended to be put into metal; but as certainly were intended to be superseded when anything better could be found. Very few machines indeed are invented which do not undergo modification. It is almost invariably the case that the second machine made has improvements on the first design. In such a machine as an Analytical Engine, this would be important; but no one ever stops a useful invention for fear of improvements. A gun or an ironclad, for example, has been scarcely made before superseded by something better, and so it must be with all inventions, though fortunately not at the same speed.

68. As to the possible modification of the engine (chapter VIII of the report) I may say that I am myself of opinion that the general design might with practical advantage be restricted. The engine would be still very useful indeed, if made not quite so automatic; even the mill by itself would, I believe, be extensively useful, if a printing or stereomoulding apparatus were joined to it. Perhaps, if that existed, the wants of further parts such as the store would be felt and supplied.

69. As to the general conclusion and recommendation (chapter IX of the report) there is little for me to notice. I see no hope of any Analytical Engine, however useful it might be, bringing any profit to its constructor, and beyond the preparation of this paper, and the publication of the volume I have mentioned as shortly to follow,

there is little or no temptation to do more. Those who wish for such an engine would, I think, give it a helping hand if they could show what pecuniary benefit it would bring. The history of Babbage's calculating machines is sufficient to damp the ardour of a dozen enthusiasts. /

SECTION C

Mechanical Notation

[250/1]

I

ON A METHOD OF EXPRESSING BY SIGNS THE ACTION OF MACHINERY

From *Philosophical Transactions of the Royal Society*, Vol. 116 (1826), pp. 250–65

Communicated 17 January, 1826, read 16 March, 1826

In the construction of an engine, on which I have now been for some time occupied, for the purpose of calculating tables and impressing the results on plates of copper, I experienced great delay and inconvenience from the difficulty of ascertaining from the drawings the state of motion or rest of any individual part at any given instant of time: and if it became necessary to enquire into the state of several parts at the same moment the labour was much increased.

In the description of machinery by means of drawings, it is only possible to represent an engine in one particular state of its action. If indeed it is very simple in its operations, a succession of drawings may be made of it in each state of its progress which will represent its whole course; but this rarely happens, and is attended with the inconvenience and expense of numerous drawings. The difficulty of retaining in the mind all the cotemporaneous and successive movements of a complicated machine, and the still greater difficulty of properly timing movements which had already been provided for, induced me to seek for some method by which I might at a glance of the eye select any particular part, and find at any given time its state of motion or rest, its relation to the motions of any other part of the machine, and if / necessary trace back the sources of its movement through all its successive stages to the original moving power. I soon felt that the forms of ordinary language were far too diffuse to admit of any expectation of removing the difficulty, and being convinced from experience of the vast power which analysis derives from the great condensation of meaning in the language it employs, I was not long in deciding that the most favourable path to pursue was to have recourse to the language of signs. It then became necessary to contrive a notation which ought if possible to be at once simple and expressive, easily understood at the commencement, and capable of being readily retained in the memory from the proper adaptation of the signs to the circumstances they were intended to represent. The first thing to be done was obviously to make an accurate enumeration of all the moving parts, and to appropriate a name to each; the multitude of different contrivances in various machinery, precluded all idea of substituting signs for these parts. They were therefore written down in succession, only observing to preserve such an order that those which jointly concur for accomplishing the effect of any separate part of the machine might be found situated near to each other: thus in a clock, those parts which belong to the striking part ought to be placed together, whilst those by which the repeating part operates ought, although kept distinct, yet to be as a whole, adjacent to the former part.

Each of these names is attached to a faint line which runs longitudinally down the page, and which may for the sake of reference be called its *indicating line*.

The next object was to connect the notation with the / drawings of the machine, in order that the two might mutually illustrate and explain each other.

It is convenient in the three representations of a machine, to employ the same letters for each part; in order to connect these with the notations, the letters which in the several drawings refer to the same parts, are placed upon the indicating lines immediately under the names of the things. If circumstances should prevent us from adhering to this rule, it would be desirable to mark those things represented in the plan by the ordinary letters of the alphabet, those pointed out for one of the other projections by the letters of an accented alphabet, and the parts delineated on the third projection by a doubly accented alphabet. In engines of so complicated a nature as to require sections at various parts as well as the three projections, this system is equally applicable, and its advantage consists in this – that the number of accents on the letter indicates at once the number of the drawing on which it appears, and when it is intended to refer to several at the same time, the requisite letters may be employed and placed in the order in which the drawings will best illustrate the part under examination.

The next circumstance which can be indicated by the system of mechanical notation which I propose, more readily than by drawings, is the number of teeth on each wheel or sector, or the number of pins or studs on any revolving barrel. A line immediately succeeding that which contains the references to the drawings is devoted to this purpose, and on each vertical line indicating any particular part of the machine, is written the number of teeth belonging to it. As there is generally a great variety of parts of machinery which / do not consist of teeth; of course every vertical line will not have a number attached to it.

The three lines immediately succeeding this, are devoted to the indication of the velocities of the several parts of the machine. The first must have, on the indicating line of all those parts which have a rectilinear motion, numbers expressing the velocity with which those parts move, and if this velocity is variable, two numbers should be written, one expressing the greatest, the other the least velocity of the part. The second line must have numbers expressing the angular velocity of all those parts which revolve; the time of revolution of some one of them being taken as the unit of the measure of angular velocity.

It sometimes happens that two wheels have the same angular velocity when they move; but from the structure of the machine, one of them rests one half of the time during which the other is in action. In this case, although their angular velocities are equal, their comparative velocities are as 1 to 2; for the second wheel makes two revolutions, whilst the other only makes one. A line is devoted to the numbers which thus arise, and it is entitled, comparative angular velocity.

The next object to be considered is the course through which the moving power is transmitted, and the particular modes by which each part derives its movement from that immediately preceding it in the order of action. The sign which I have chosen to indicate this transmission of motion (an arrow), is one very generally employed to denote the direction of motion in mechanical drawings; it will therefore readily suggest the direction in which the movement is transmitted. There are however various ways by which / motion is communicated; and it becomes a matter of some importance to consider whether, without interfering with the sign just selected, some modification might not be introduced into its minor parts, which, leaving it unaltered in the general form, should yet indicate the peculiar nature of the means by which the movement is accomplished.

On enumerating those modes in which motion is usually communicated, it appeared that they may be reduced to the following.

One piece may receive its motion from another by being permanently attached to it as a pin on a wheel, or a wheel and pinion on the same axis.	This may be indicated by an arrow with a bar at the end. ┼——→
One piece may be driven by another in such a manner that when the driver moves, the other also always moves; as happens when a wheel is driven by a pinion.	An arrow without any bar. ———→
One thing may be attached to another by stiff friction.	An arrow formed of a line interrupted by dots. —·—·—·→
One piece may be driven by another, and yet not always move when the latter moves; as is the case when a stud lifts a bolt once in the course of its revolution.	By an arrow, the first half of which is a full line, and the second half a dotted one. ———·····→
One wheel may be connected with another by a ratchet, as the great wheel of a clock is attached to the fusee.	By a dotted arrow with a ratchet tooth at its end. ·⋀·········→

Each of the vertical lines, representing any part of the machine, must now be connected with that representing the part from which it receives its movement, by an arrow of such a kind as the preceding table indicates; and if any part derives motion from two or more sources, it must be connected by the proper arrows with each origin of its movement. It will in some cases contribute to the better understanding of the machine, if those parts which derive movement from two or more sources have their names connected by a bracket, with two or more vertical lines, which may be employed to indicate the different motions separately. Thus if a shaft has a circular as well as a longitudinal motion, the two lines attached to its name should be characterised by a distiguishing mark, such as (vertical motion) and (circular motion). Whenever any two or more motions take place at the same time, this is essential; and when they do not, it is convenient for the purpose of distinguishing them.

All machines require, after their parts are finished and put together, certain alterations which are called adjustments. Some of these are permanent, and, when once fixed by the maker, require no further care. Others depend on the nature of the work they are intended to perform, as in the instance of a corn mill; the distance between the stones is altered according to the fineness of the flour to be ground: these may be called usual adjustments; whilst there are others depending on the winding up of a weight, or spring, which may be called periodic adjustments. As it is very desirable to know all the adjustments of a machine, a space is reserved, below that in which the connections of the moving parts are exhibited, where these may be indicated; if there are many / adjustments, this space may be subdivided into three, and appropriated to each of the three species just enumerated, the permanent, the usual, and the periodic: if their number is small, it is better merely to distinguish them by a modification of their signs. It is sometimes impossible to perform such adjustment, except in a particular succession; and it is always convenient to adhere to one particular order. Numbers attached to their respective lines denote the order in which the parts are to be adjusted; and as it will sometimes happen that two or more adjustments must be made at the same time, in that case the same numbers must be written on the lines belonging to all those parts which it is necessary to adjust simultaneously.

If it is convenient to distinguish between the species of adjustments without separating them by lines, this may be accomplished by putting a line above or below the figures, or enclosing them in a circle, or by some similar mode. I have attached the letter P to those which are periodic.

It would add to the knowledge thus conveyed, if the sign indicating adjustment also gave us some information respecting its nature; there are, however, so many different species, that it is perhaps better in the first instance to confine ourselves to a few of the most common, and to leave to those whom may have occasion to employ this kind of notation, the contrivance of signs, fitted for their more immediate purpose.

One of the most common adjustments is that of determining the distance between two parts, as between the point of suspension and the centre of oscillation of a pendulum. This might be indicated by a small line crossing the vertical / line attached to the part.

If the distance between two parts, which are represented by different lines, is to be altered, their lines may be connected by a horizontal line.

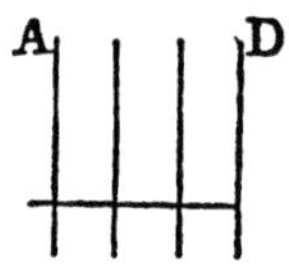

Thus the adjustment of A in the above figure depends on its distance from D.

This adjustment often determines the length of a stroke, and sometimes the eccentricity of an eccentric, which depends on the linear distance between its centre and its centre of motion.

The next most frequently occurring adjustment is that which is sometimes necessary in fixing two wheels, or a wheel and an arm on the same axis. This relation of angular position, may be indicated by a circle and two radii placed at the requisite angle, or that angle may be stated in figures and enclosed within the circle; this circle ought however to be connected by a line with the other part, with which the angle is to be formed, thus,

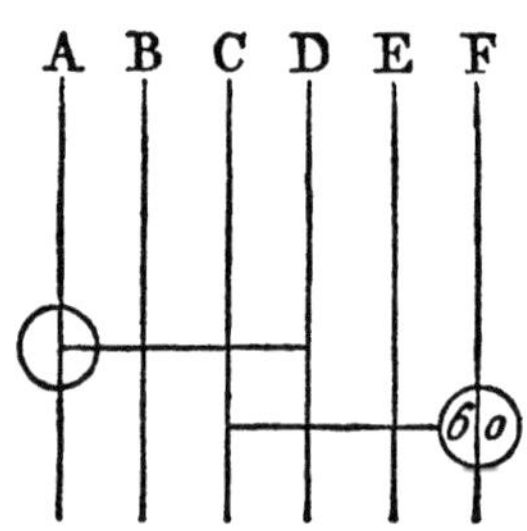

which means that an adjustment is to be made by fixing A on its axis, making a right angle with D, and that F must also make with C an angle of sixty degrees. In speaking of these angles, it should always be observed that they refer to the angles made by the parts on one of the planes of projection. /

When it is thought requisite to enter into this minute detail of adjustments, it will be necessary, in order to avoid confusion, to put the lines indicating the order of adjustment, above and distinct from these signs.

The last and most essential circumstance to be represented, is the succession of the movements which take place in the working of the machine. Almost all machinery, after a certain number of successive operations, recommences the same course which it had just completed and the work which it performs usually consists of a multitude of repetitions of the same course of particular motions.

It is one of the great objects of the notation I am now explaining, to point out a method by which, at any instant of time in this course or cycle of operations of any machine, we may know the state of motion or rest of every particular part; to present a picture by which we may, on inspection, see not only the motion at that moment of time, but the whole history of its movements, as well as that of all the contemporaneous changes from the beginning of the cycle.

In order to accomplish this, each of the vertical indicating lines representing any part of the machine, has, adjacent to it, other lines drawn in the same direction: these accompanying lines denote the state of motion or rest of the part to which they refer, according to the following rules.

1. Unbroken lines indicate motion.
2. Lines on the right side indicate that the motion is from right to left.
3. Lines on the left side indicate that the direction of the motion is from left to right. /
4. If the movements are such as not to admit of this distinction, then when lines are drawn adjacent to an indicating line, and on opposite sides of it, they signify motions in opposite directions.
5. Parallel straight lines denote uniform motion.
6. Curved lines denote a variable velocity. It is convenient as far as possible to make the ordinates of the curve proportional to the different velocities.
7. If the motion may be greater or less within certain limits:

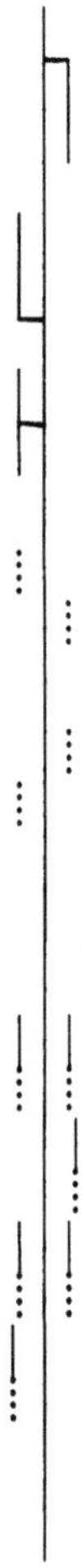

then if the motion begin at a fixed moment of time, and it is uncertain when it will terminate, the line denoting motion must extend from one limit to the other, and must be connected by a small cross line at its commencement with the indicating line. If the beginning of its motion is uncertain, but its end determined, then the cross line must be at its termination. If the commencement and the termination of any motion are both uncertain, the line representing motion must be connected with the indicating line in the middle by a cross line.

8. Dotted lines imply rest. It is convenient sometimes to denote a state of rest by the absence of any line whatever.

9. If the thing indicated be a click, bolt, or valve, its dotted line should be on the right side if it is out of action, unbolted, or open, and on the left side if the reverse is the case. /

10. If a bolt may rest in three positions: 1st, bolted on the right side; 2nd, unbolted; 3rd, bolted on the left side. When it is unbolted, and in the middle station, use two lines whilst in the act of unbolting, and two lines of dots, one on each side of, and close to the indicating line, whilst it rests in this position. When it is bolted on the right side, a line or a line of dots at a greater distance on the right hand from the indicating line will represent it. And if it is bolted on the left, a similar mode of denoting it must be used on that side. Any explanation may, if required, be put in words at the end of the notation, as will be observed in that of the hydraulic ram, Plate IV.

I have now explained means of denoting by signs almost all those circumstances which usually occur in the motion of machinery: if other modifications of movement should present themselves, it will not be difficult for anyone who has rendered himself familiar with the symbols employed in this paper, to contrive others adapted to the new combinations which may present themselves.

The two machines which I have selected as illustrations of the application of this method, are, the common eight-day clock, and the hydraulic ram. The former was made choice of from its construction being very generally known, and I was induced to choose the latter from the apparent difficulty of applying this method to its operations.

The advantages which appear to result from the employment of this *mechanical notation*, are to render the description of machinery considerably shorter than it can be when expressed / in words. The signs, if they have been properly chosen, and if they should be generally adopted, will form as it were a universal language; and to those who become skilful in their use, they will supply the means of writing down at sight even the most complicated machine, and of understanding the order and succession of the movements of any engine of which they possess the drawings and the mechanical notation. In contriving machinery, in which it is necessary that numerous wheels and levers, deriving their motion from distant parts of the engine, should concur at some instant of time, or in some precise order, for the proper performance of a particular operation, it furnishes most important assistance; and I have myself experienced the advantages of its application to my own calculating engine, when all other methods appeared nearly hopeless.

DESCRIPTION OF THE PLATES[a]

Plates I and II are different representations of an eight-day clock; for the purpose of comparing it with the notation.

Plate III represents the mechanical notation of the same clock:

Under the names of each part follow the letters which distinguish them in the plates.

The next line contains numbers which mark the number of teeth in each wheel, pinion, or sector.

The following line is intended to contain numbers expressing the linear velocity of the different parts: in the eight-day clock this line is vacant, because almost all the motions are circular.

The next line indicates the angular velocity of each part; / and in order to render the velocity of the striking parts comparable with those of the time part, I have supposed one revolution of the striking fusee to be made in one minute: I have also taken one revolution of the scapement wheel, or one minute of time as the unit of angular velocity.

The space entitled comparative angular velocity, expresses the number of revolutions one wheel makes during one revolution of some other; it differs

[a] The plates, which have been renumbered, follow this article on p. 221.

from the real angular velocity, because one wheel may be at rest during part of the time. Thus the clock strikes 78 strokes during twelve hours, or one revolution of the hour hand: if this be called unity, the pin wheel moves through 78 pins or 9¾ revolutions in the same time; its comparative angular velocity is therefore 9¾.

The space in which the origin of motion is given, will not require any explanation after reading the description of the signs employed in this paper.

The adjustments are numbered in the order in which they are to be made. No. 1, is attached to the crutch: the first adjustment is to set the clock in beat. No. 2, is to adjust the length of the pendulum to beat seconds. No. 3, occurs in three different places, at the hour and minute hands, and at the snail on the hour wheel. It is necessary that when the hour hand is at a given figure, three o'clock for instance, that the minute hand should be set to twelve o'clock; it is also necessary that the snail should be in such a position that the clock may strike three. These adjustments must be made at the same time. No. 4, is for the adjustment of the seconds hand to 60 seconds. No. 5, is double, and is for the adjustment of the minute and hour hand to the next whole / minute to that which is indicated by the watch by which the clock is set. No. 6, is for the pendulum, which must be held aside at its extreme arc until the instant at which the watch reaches the time set on the face of the clock; it must then be set free.

The remaining part of the notation indicates the action of every part at all times; but as the whole cycle of twelve hours would occupy too much space, a portion only is given about the hour of four: from this the machine may be sufficiently understood. As an instance of its use, let us enquire what movements are taking place at seven seconds after four o'clock. On looking down at the left-hand side to the time just mentioned, we observe between the end of the sixth, and end of the seventh second, that the pendulum and crutch begin to move from the right to the left, increasing their velocity to a maximum, and then diminishing it; that the whole train of wheels of the time part are at rest during the greater part of that second, and all move simultaneously a little before its termination. The greater part of the train of the striking part is moving uniformly; but two parts the crosspiece, and the other moving the hammer, being at the commencement of this second in a state of motion from right to left, suddenly have that motion reversed for a short time: this is at the moment of striking: two other pieces, the hawk's bill, and the gathering pallet, appears to act at the same moment.

If the course of movement of any one part is required throughout the whole cycle of the machine's action, we have only to follow its indicating line. If it is required to find what motions take place at the same time, we have only to look along the horizontal line marked by the time specified. /

Let us now enquire into the source of motion of the minute hand. On looking down to the space in which the origin of motion is given, we observe an arrow point, which conveys us to the:

Cannon pinion, with which it is connected permanently.

The cannon pinion is driven by the centre or hour wheel, with which it is connected by stiff friction.

The hour wheel is driven by its pinion, to which it is permanently attached.

The hour wheel pinion is driven by the great wheel, into which it works.

The great wheel is driven by the fusee, with which it is connected by a ratchet.

The fusee is driven by the spring barrel or main spring, which is the origin of all the movements.

When that part of the notation which relates to the successive movements of the machine is of considerable extent, it is convenient to write on a separate piece of paper the names of every part, at the same distances from each other as the indicating lines, and exactly as they are placed at the top. By sliding this paper down the page to any part which is under consideration, the trouble of continual reference to the top of the drawing will be avoided.

Plate IV represents the hydraulic ram: its mechanical notation is added below it.

A is the supplying pipe.
B is the great valve.
C the valve into the air vessel.
D the air vessel.
E the ascending water.
F the small air valve – its office is to supply a small / quantity of air at each stroke; it opens when the valve *C* is just closed, and a regurgitation takes place in the supplying pipe just previous to the opening of the great valve. Without this contrivance, the pressure on the air in the air vessel would cause it to be soon absorbed by the water, and the engine would cease to act.

In this notation two indicating lines *A*, *A*, are allowed to the supplying water, because it takes three different courses during the action of the machine. The first of these marks the time of its motion when it enters the air vessel, and the second indicates its course when passing through the great valve, and also its course when, owing to the elasticity of the materials, its motion is for an instant reversed, at which moment air is taken in at the air valve *F*.

The action of the machine is as follows: the supplying water rushing along the great pipe passes out at the great valve; it acquires velocity until the pressure of the effluent water against the under part of the great valve causes it to close suddenly. At this moment the whole momentum of the water is directed against the sides of the machine, and the air valve being the weakest part gives way, and admits a small quantity of water; the air spring soon resists sufficiently to close the air valve: at this moment the elasticity of the apparatus reacting on the water in the great pipe, drives it back for an instant, during which the pressure of the atmosphere opens the air valve, and a small quantity of air enters; this finds its way to the air chambers, which easily discharges it through the ascending pipe if too much air has entered. /

PLATE I

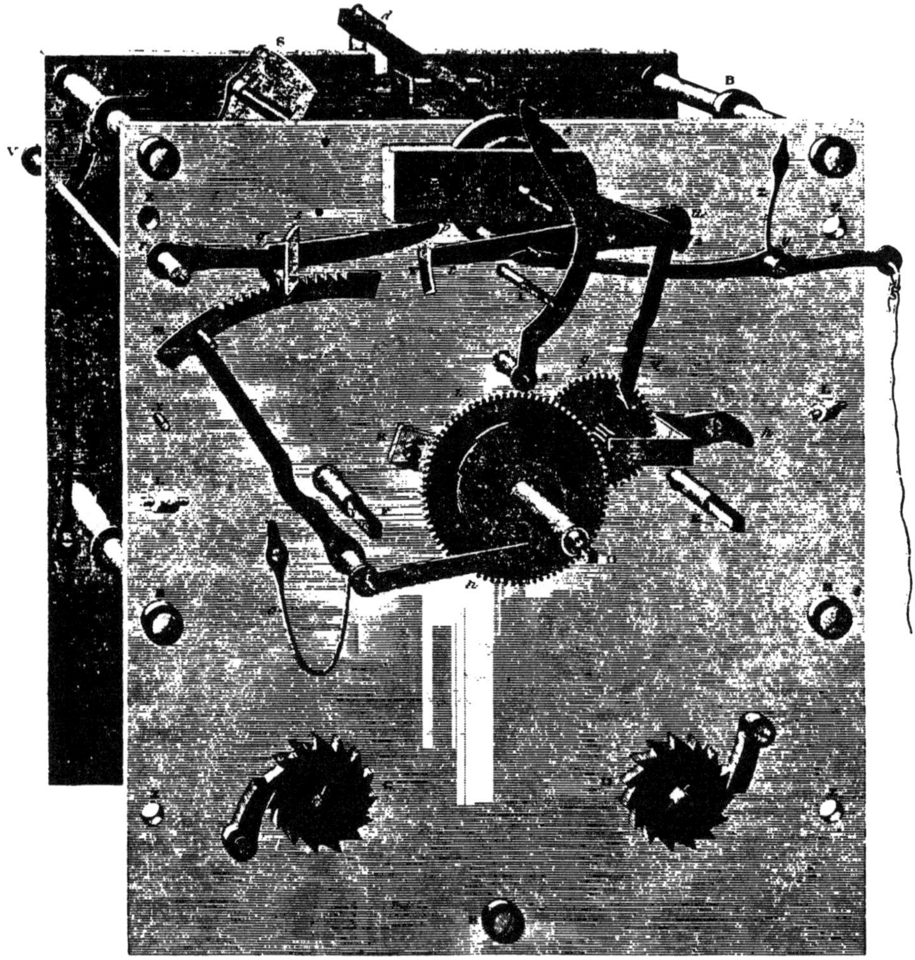

PLATE II

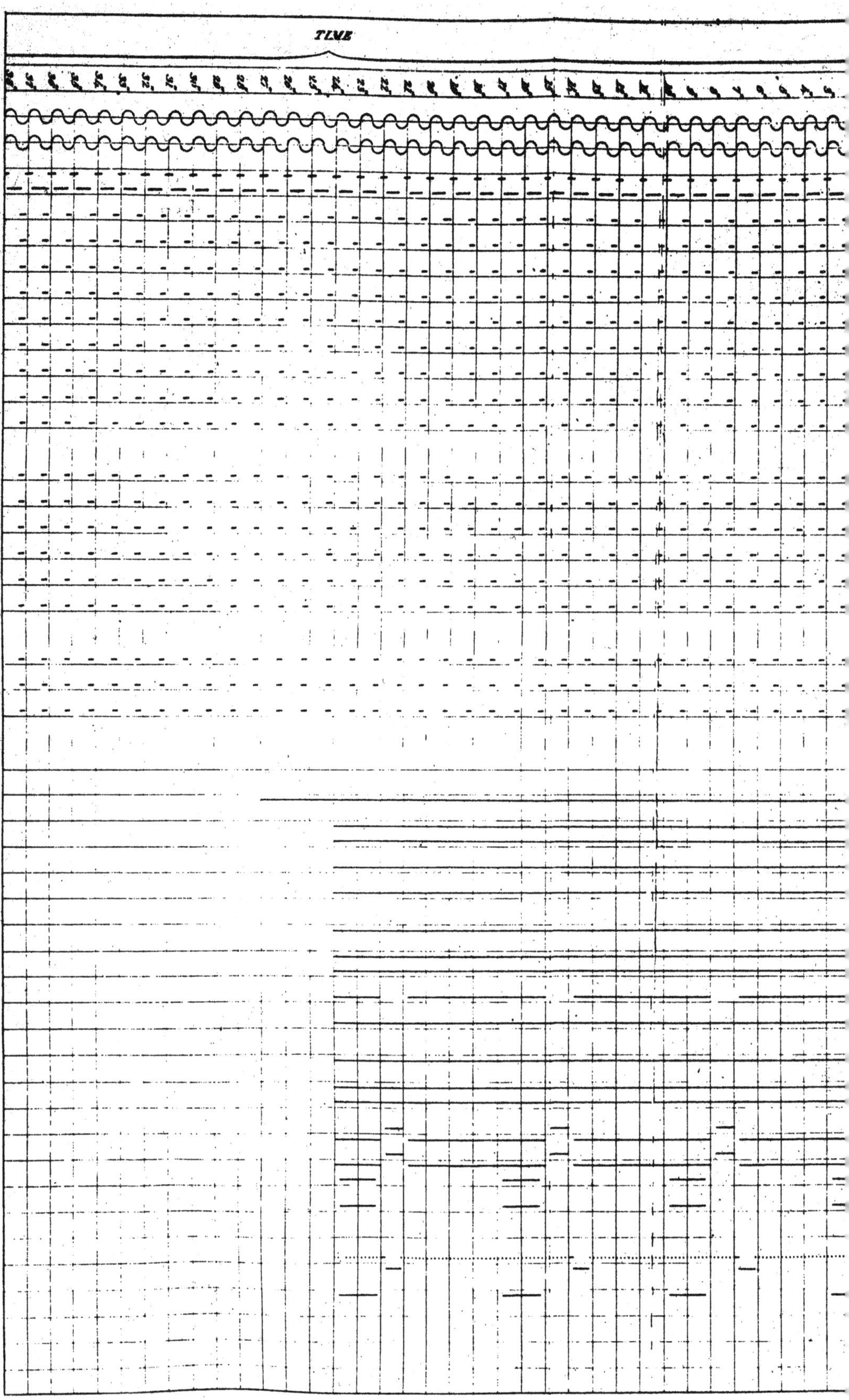

E. Turrell. sc.

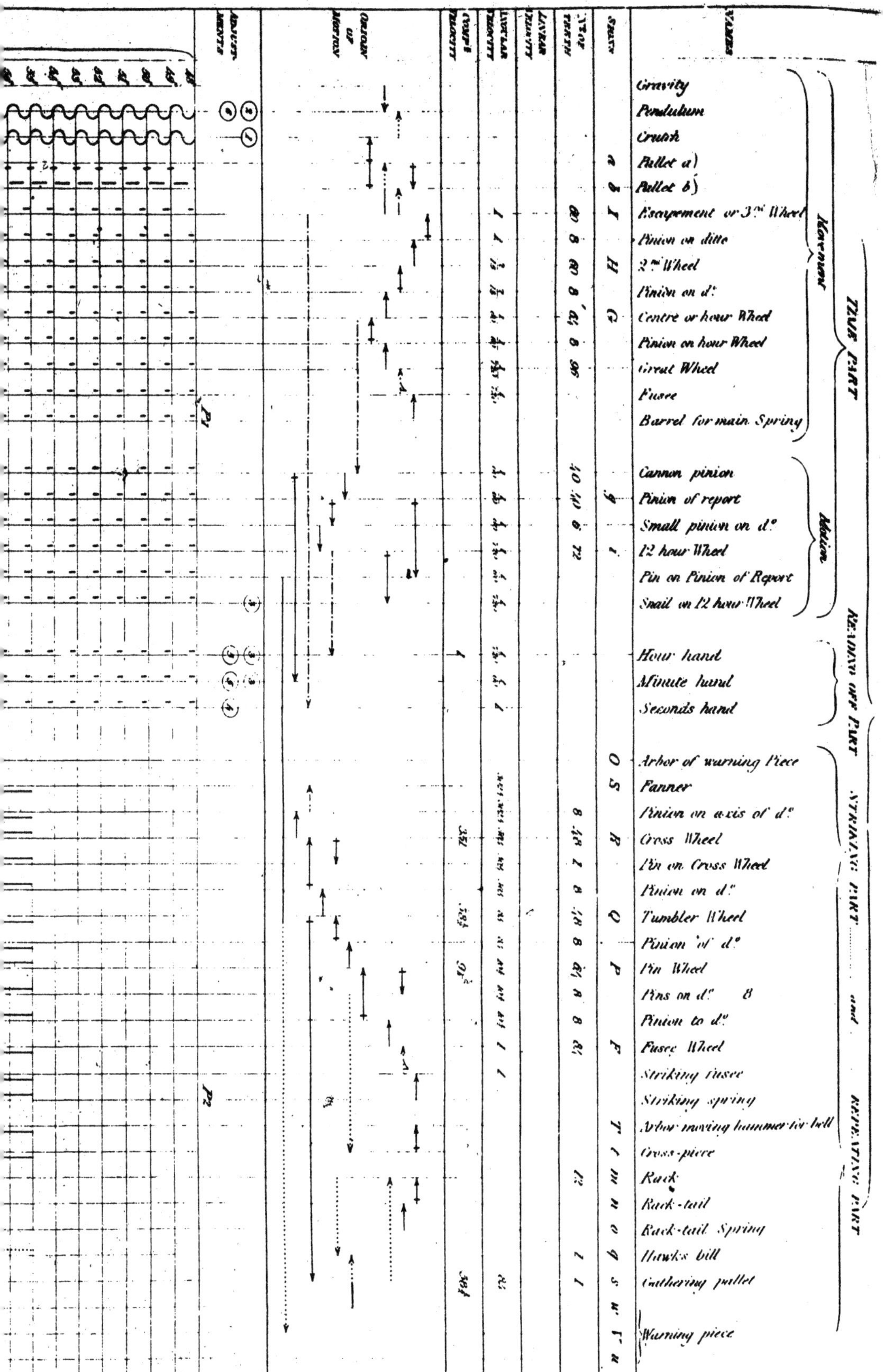

NAMES
Gravity
Pendulum
Crutch
Pallet a)
Pallet b)
Escapement or 3rd Wheel
Pinion on ditto
2nd Wheel
Pinion on d°
Centre or hour Wheel
Pinion on hour Wheel
Great Wheel
Fusee
Barrel for main Spring
Movement
Cannon pinion
Pinion of report
Small pinion on d°
12 hour Wheel
Pin on Pinion of Report
Snail on 12 hour Wheel
Motion
Hour hand
Minute hand
Seconds hand
Arbor of warning Piece
Fanner
Pinion on axis of d°
Cross Wheel
Pin on Cross Wheel
Pinion on d°
Tumbler Wheel
Pinion of d°
Pin Wheel
Pins on d° 8
Pinion to d°
Fusee Wheel
Striking fusee
Striking spring
Arbor moving hammer for bell
Cross-piece
Rack
Rack-tail
Rack-tail Spring
Hawks bill
Gathering pallet
Warning piece
TIME PART
READING OFF PART
STRIKING PART and REPEATING PART
P1
P2

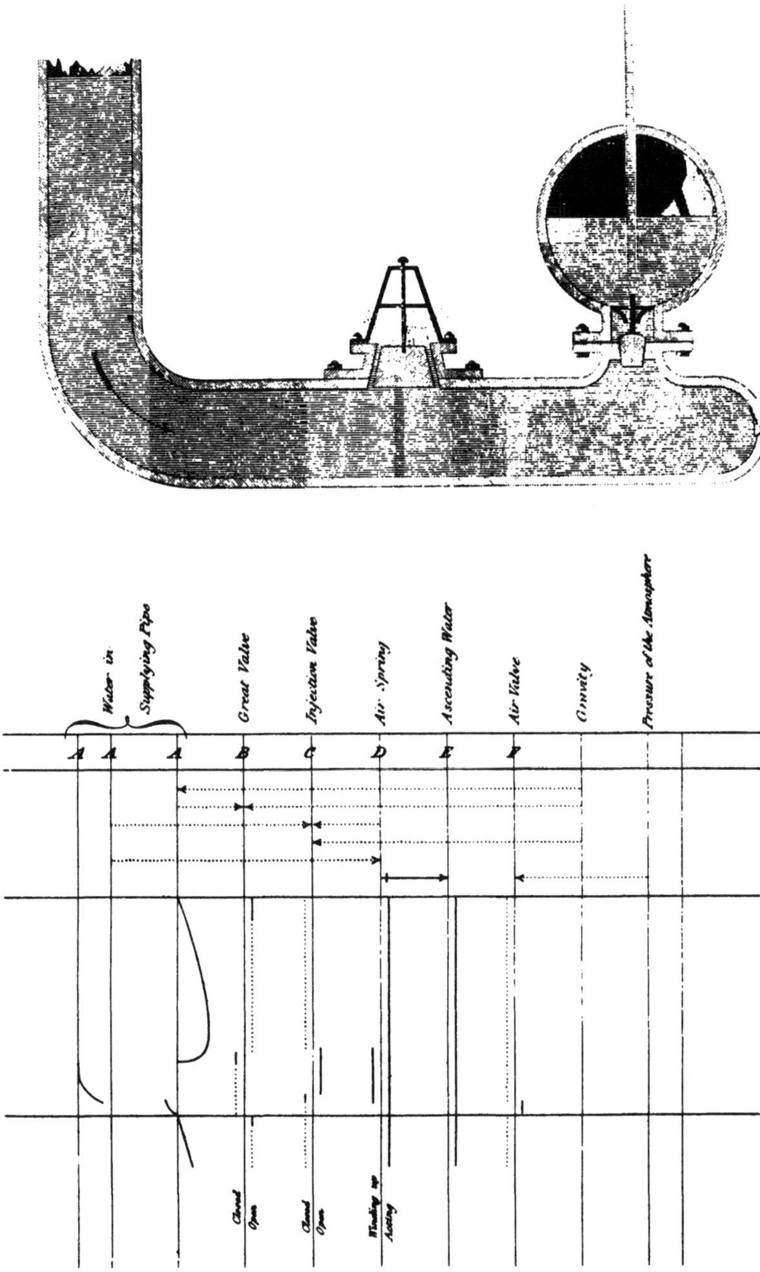

Water in
Supplying Pipe
Great Valve
Injection Valve
Air Spring
Ascending Water
Air Valve
Gravity
Pressure of the Atmosphere
A
A
A
B
C
D
E
F
Closed
Open
Closed
Open
Winding up
Acting

PLATE IV

[1]

II

LAWS OF MECHANICAL NOTATION

(London, 1851)

CHAP. I.

ON LETTERING DRAWINGS

All machinery consists of—

Framing		Parts or Pieces
Fixed	Movable	Movable as axes, springs, etc.

Every *piece* possesses one or more *working points.* These are divided into two classes, those by which the piece acts on others, and those by which it receives action from them: these are called *driving* and *driven points.* A *working point* may fulfil both these offices, as, for example, the same teeth which are driven by one wheel, may in another part of their course drive other wheels.

The following alphabets of large letters are used in drawings:

Etruscan		Roman		Writing	
A	*A*	A	*A*	A	*A*
B	*B*	B	*B*	B	*B*
C	*C*	C	*C*	C	*C*
. . .		. . .		. . .	

The following alphabets of small letters are used:

a b c d...
a b c d...

It is most convenient, and generally sufficient, to use only the letters *a, c, e, i, m, n, o, r, s, u, v, w, x, z,* of both these latter alphabets.

Rule 1. Every separate portion of *framework* must be indicated by a large *upright* letter.

Rule 2. Every *working point* of framework must be indicated by a *small* printed letter. /

Rule 3. *Framework* which is itself movable, must be represented by a *large upright* letter, with the sign of motion in its proper place below it (*see signs of motion*), as

$$\underset{-}{G} \qquad \underset{\smile}{H}$$

Rule 4. In lettering drawings, commence with the axes. These must be lettered with *large inclined* letters, of either of the three alphabets. Whenever the wheels or arms of any two or more adjacent axes cross each other on the plan, avoid denoting those axes by letters of the *same* alphabet.

Rule 5. In lettering *pieces* as wheels, arms, etc., belonging to any axis, whether they are fixed to it or movable upon it, always use *inclined capitals* of the *same* alphabet as that of the letter representing the axis.

Rule 6. Beginning with the lowest *piece* upon an axis, assign to it any *capital* letter of the *same* alphabet. To the piece next above, assign any other *capital* letter which occurs *later* in the *same* alphabet. Continue this process for each *piece*.

Thus, although the succession of the letters of the *same* alphabet need not be continuous, yet their occurrence in *alphabetic* order will never be violated.

Rule 7. In lettering *pieces* upon axes perpendicular to the elevation, or to the end views, looking from the left side, the earliest letters of the alphabet must be placed on the pieces most remote from the eye.

Rule 8. No axis, which has a *piece* crossing any other *piece* belonging to an adjacent axis, must have the same identity as that axis.

If there are many *pieces* on the same axis, it may be necessary to commence with one of the earlier letters of the alphabet.

Rule 9. In placing letters representing any *piece* on which portions of other *pieces* are projected, it is always desirable to select such a situation that no doubt can be entertained as to which of those *pieces* the letter is intended to indicate. This can often be accomplished by placing the letter upon some portion of its own *piece* which extends beyond the projected parts of the other *piece*.

Rule 10. When *pieces* are very small, or when they are crossed by many other lines, it is convenient to place the letter representing them outside the *piece* itself, and to connnect it with the *piece* it indicates by an arrow. This arrow should be a short fine line terminated by a head, abutting on, or perhaps projecting into, the piece represented by the letter.

Rule 11. When upon any drawing, a letter having a dot beneath it occurs, it marks the existence of a *piece* below. /

Rule 12. In case another *piece*, exactly similar to one already represented and lettered, exists below it, it cannot be expressed by any visible line. It may, however, be indicated by placing its proper letter outside, and connecting that letter with a *dotted* arrow abutting on the upper *piece*.

Rule 13. The permanent connection of two pieces of matter, or the permanent gearing of two wheels, is indicated by a short line, crossing, at right angles, the point of contact. The sign | indicates, in a certain sense, fixed connection. This sign will be found very useful for indicating the boundaries of various pieces of framing.

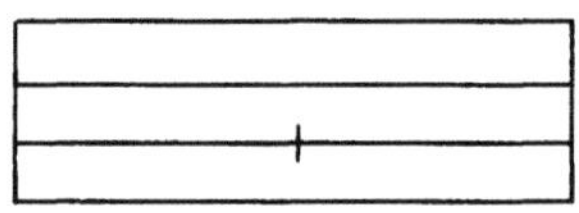

Of the indices of letters

Rule 14. Various indices and signs may be affixed to letters. Their position and use are indicated in the subjoined letter:

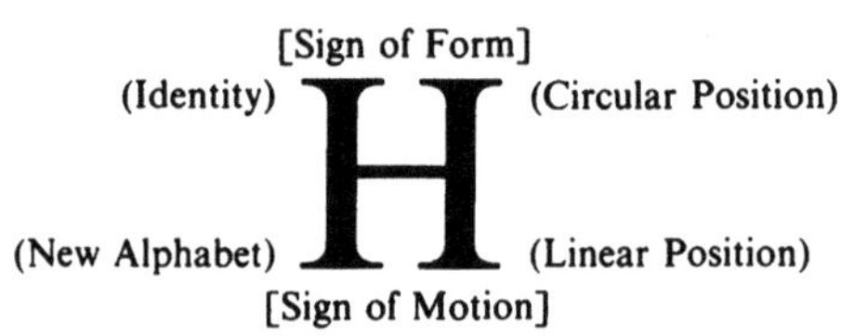

Rule 15. The index on the left-hand upper corner is used to mark the identity of two or more parts of a *piece* which are permanently united; each being denoted by a letter with the *same* index.

Rule 16. It is used also to connect any *piece* itself with its various working points. Thus all the small letters which indicate the working points, must have the *same* index of identity as the letter expressing the *piece* itself.

Rule 17. Every *working point* must be marked by the *same small letter* as the *working point* of the *piece* upon which it acts.

Rule 18. The bearings in which axes work, as well as the working surface of the axes themselves, and also the working surfaces of slides, are *working points*, and must be lettered as such.

Of the index of linear position

The successive order in which the various *pieces* upon one axis succeed each other is indicated by the alphabetic succession. /

It may, however, in some cases be convenient to distinguish between the relative heights of the various arms or wheels which constitute one *piece*.

This may be easily accomplished by means of the index of *linear position*.

Every piece may be represented as a whole, by one letter, with its proper index of identity. If, however, it is necessary to distinguish the different arms or parts of which it is composed, so as to indicate their relative position above the plane of projection, this may be accomplished by means of the indices of linear position.

Rule 19. If ${}^{3}P$ represent the whole of any *piece*, ${}^{3}P_1$, ${}^{2}P_2$, ${}^{3}P_3$, ${}^{3}P_4$, etc., will represent in succession the several arms or parts of which ${}^{3}P$ is composed: ${}^{3}P_1$ indicating that which is most distant from the eye.

Of the index of circular position

It may occasionally be desirable to indicate the order of succession in angular position of the various arms belonging to the same *piece*, when projected on a plane. The index on the right hand upper corner is applied to this purpose.

Rule 20. ${}^{6}R$ representing any *piece*,

${}^{6}R^1$ will represent any arm as the origin,

${}^{6}R^2$ the next arm in angular position in the direction '*screw*', that is, from left to right,

${}^{6}R^3$ the next, etc.

Thus,

$${}^{6}R^1_1,\ {}^{6}R^2_2,\ {}^{6}R^3_3 \ldots {}^{6}R^n_n$$

would represent n arms placed spirally round an axis at various heights above it.

Of the index of new alphabet

In case the three alphabets given above are found insufficient, the index on the left lower side is reserved to mark new alphabets. In the most complicated drawing I have scarcely ever had occasion to use it. It might in some cases be desirable to have a fourth alphabet, differing in form from those already given.

MR BABBAGE *will feel obliged by any criticisms, or additions to these rules of drawing, and to the mechanical alphabet, and requests they may be addressed to him by post, at No.* 1, *Dorset Street, Manchester Square.*

July 1851

III

NOTE SUR LA MACHINE SUÉDOISE DE MM. SCHEUTZ POUR CALCULER LES TABLES MATHÉMATIQUES PAR LA MÉTHODE DES DIFFÉRENCES ET EN IMPRIMER LES RÉSULTATS SUR DES PLANCHES STÉRÉOTYPES

Académie des sciences, Paris, Comptes-rendus hebdomadaires des séances, Vol. 41, (1855), pp. 557–60

Pour faire connaître plus exactement le mode de construction ou la théorie de l'admirable instrument exposé maintenant par MM. Scheutz dans le Palais de l'Industrie, j'ai l'honneur de mettre sous les yeux de l'Académie, d'après le désir qui m'en a été exprimé par quelques Académiciens, une série de tableaux graphiques. Ces tableaux, exécutés par mon fils M. Henry Prevost Babbage, offrent une application à la machine suédoise du système de notation employé par moi, depuis un assez grand nombre d'années déjà, pour la description et l'explication des machines.

La *notation mécanique* comprend plusieurs sections dont voici les principales:

1°. Les principes de l'emploi des lettres sur les dessins des machines;
2°. L'alphabet des formes;
3°. L'alphabet des mouvements;
4°. Les trains;
5°. Les cycles;
6°. Les notations d'opération;
7°. Les notations analytiques. /

La forme exacte d'une machine se connaît en général par la vue des dessins. Dans un grand nombre de circonstances néanmoins, ce moyen manque de la précision désirable.

[a] An English translation of this paper appears as the next article in this volume.

Jusqu'ici l'emploi des lettres marquant les différentes parties d'un dessin a été presque entièrement arbitraire. Dans la première section de la *notation mécanique* sont posées les règles d'après lesquelles les lettres elles-mêmes deviennent directment des signes, et donnent des indications importantes. Il est alors presque inutile de se reporter du plan à l'élévation. Voici quelques-unes de ces règles:

Les bâtis ou supports (framework) s'expriment par des capitales droites ou des minuscules droites; les pièces mobiles par des capitales penchées ou par des minuscules penchées.

Framework	*Pièces mobiles*
A, B, C, a, b, c,	*A, B, C, a, b, c.*

L'avantage de cette règle, c'est que si vous examinez un dessin dans un but d'étude, pour vous rendre compte de la destination ou du jeu de tel ou tel organe; au lieu de fatiguer votre attention, vous pouvez faire entièrement abstraction de toutes les parties qui se rapportent seulement au *framework*.

Pour comprendre parfaitement une machine, il est indispensable d'en connaître avec exactitude:

1°. La forme;

2°. Les différents récepteurs par lesquels les mouvements sont transmis depuis le premier moteur jusqu'au résultat final: c'est ce que j'appelle les *trains*;

3°. Pour les diverses parties de la machine, le temps précis où commence le mouvement, et la durée de ce mouvement: c'est là le *cycle*.

Une machine est composée de pièces fixes et de pièces mobiles, de pièces qui impriment le mouvement ou qui le reçoivent. Chacune de ces pièces peut être considérée isolément, c'est-à-dire en elle-même. Les points ou les surfaces par lesquels chaque pièce imprime le mouvement ou le reçoit, sont les parties principales de cette pièce. Ces points s'appellent *opérateurs* ou *working points*—de là une distinction importante à observer dans le choix des lettres.

Pour marquer les rapports des *points opérateurs* de chaque pièce avec les grandes lettres qui les représentent, on a donné à chaque grande lettre un indice supérieur, du côté gauche. C'est l'*indice d'identité.* /

Les petites lettres représentant un des points opérateurs sont marquées respectivement du même signe d'identité; ainsi *a*, *b*, *c*, etc., sur un dessin, appartiennent à une lettre capitale adjacente, ^{3}D par exemple, seront écrits: ^{3}a, ^{3}b, ^{3}c, etc.

Les axes, comme les pièces glissantes, pourront être désignés par les grandes lettres de cinq ou six alphabets différents. Etrusque, **A**, **B**, **C**; romain, A, B, C; italique, $\mathscr{A}$, $\mathscr{B}$, $\mathscr{C}$, etc.

Mais quand l'alphabet a été déterminé, les pièces quelconques, soit fixes, soit mobiles, se rapportant à cet axe, doivent être représentées par une lettre capitale de l'alphabet attribué à l'axe lui-même.

A l'aide de ce moyen, on peut distinguer aisément, sur le dessin le plus chargé, l'axe auquel telle ou telle pièce appartient. En effet, ces caractères de l'alphabet affectés aux axes doivent être choisis de telle sorte que deux axes dont les pièces se croisent ne soient jamais marqués du même caractère.

L'exécution des dessins de la *machine analytique* n'a présenté que très-peu de cas par lesquels il eût été désirable de se servir de quatre ou même d'un plus grand nombre des alphabets différents. ou même d'un plus grand nombre. En réalité, il n'en a pas été employé plus de trois.

La position relative des pièces, soit fixes, soit mobiles, appartenant à des axes ou à des pièces glissantes, peut toujours se voir sur chaque projection si l'on adopte la règle ci-après.

Commencer à noter les lettres sur la pièce la plus éloignée appartenant à l'axe, en prenant une lettre convenable; ensuite, et par ordre de proximité, marquer chaque autre pièce des lettres qui, dans l'alphabet, viennent plus bas qu'aucune de celles dont on s'est servi pour les parties précédentes.

Voici, en peu de mots, l'application de ce système de notation à la description de la machine suédoise.

Dans la série des tableaux graphiques exposés maintenant sous les yeux de l'Académie, les seules sections du système de notation employées par M. Henry P. Babbage sont les *lettres*, les *trains*, les *cycles*.

Le principal tableau A, long de plus de 4 mètres, est relatif au cycle. Il contient environ 700 colonnes verticales. Le *temps du mouvement* de chaque pièce, et le *temps de l'action* de chacun de ses *points opérateurs*, peut immédiatement se voir avec la plus grande facilité.

Les lettres rouges, au haut des colonnes, marquent la liaison des *cycles* avec les *trains*. /

Le *cycle*, ou la période de la machine, emploie 96 unités de temps. Pendant le cycle, un calcul se trouve complétement exécuté et imprimé.

Sur le tableau B sont distinctement indiquées, au nombre de plusieurs centaines, toutes les pièces mobiles de la machine.

Chacune de ces pièces montre ses *points opérateurs*. On voit de quelle manière elle reçoit les mouvements, et comment elle les transmet à d'autres pièces.

Le signe de la manivelle, à gauche des *trains*, est le point de départ du mouvement. Après s'être subdivisé, et recomposé sous une grande variété de formes, ce mouvement vient s'arrêter au côté droit des trans par le résultat collectif du jeu de toutes les parties de la machine.

Une série de poinçons d'acier portant des caractères numériques disposés par le mécanisme d'après la loi de la table qu'il s'agit de calculer, se présente successivement, ligne par ligne, au-dessus d'une plaque de plomb. Cette plaque, après chaque opération, vient se presser contre les poinçons et reçoit ainsi, sans possibilité d'erreur, l'ampreinte des chiffres qui sont le résultat définitif du calcul. /

IV

NOTE ON THE SWEDISH MACHINE OF MESSRS SCHEUTZ TO CALCULATE MATHEMATICAL TABLES BY THE METHOD OF DIFFERENCES AND TO PRINT THE RESULTS ON STEREOTYPE PLATES[a]

Translated from *Académie des sciences, Paris, Comptes-rendus hebdomadaires des séances*, Vol. 41 (1855), pp. 557–60

In order to convey more exactly the method of construction or the theory of the admirable instrument now being exhibited by Messrs Scheutz at the Palais de l'Industrie, I have the honour of submitting before the *Académie*, in accordance with the wish expressed to me by several of its members, a series of diagrams. These diagrams, produced by my son Mr Henry Prevost Babbage, provide an application to the Swedish machine of the system of notation used by me for some years already for the description and explanation of machines.

The system of *Mechanical Notation* is made up of many sections of which the following are the main ones:

1. The principles of lettering on machine drawings;
2. The alphabet of form;
3. The alphabet of motion;
4. The trains;
5. The cycles;
6. The operational notation;
7. The analytical notation.

The exact form of a machine is generally clear from seeing its drawing. Nevertheless in many cases this method lacks the necessary precision.

---

[a] This paper is a new English translation of the preceding article in this volume.

Until now, the use of letters to mark the different parts of a diagram has been almost entirely arbitrary. In the first section of the *Mechanical Notation* rules are laid down under which the letters themselves provide important information. It is then almost unnecessary to refer from the plan to the elevation. Here are a few of these rules:

The frames or supports (Framework) are indicated by capital or small upright letters; the moving parts by capital or small inclined letters.

Framework	*Moving pieces*
A, B, C, a, b, c,	*A*, *B*, *C*, *a*, *b*, *c*.

The advantage of this rule is that, if you examine a drawing with a particular purpose in mind, to find out the motion or the action in a particular part; instead of wasting your concentration, you may entirely discount all the components which are only part of the *framework*.

To understand a machine perfectly one must know accurately:

1. Its form;

2. The different driven parts by which its movements are transmitted from the first mover to the final result: these are what I call the *trains*;

3. For the different parts of the machine the exact time at which movement begins, and the duration of the movement: this is the *cycle*.

A machine is made up of fixed and moving parts, of parts which transmit movement or parts which are moved. Each of these can be considered separately, that is to say, by itself. The points or the surfaces by which each piece acts or is acted upon, are its principal parts. These points are called *operational* or *working points*. As a result of this there is an important distinction to be observed in the choice of letters.

To record the relationships between the *working points* of each piece, and the capital letters which represent the piece, each capital has a superscript number on its left. This is the *index of identity*.

The small letters which represent working points are each marked with the same index of identity; thus *a*, *b*, *c*, etc., on a diagram, which belong to an adjacent capital, ^3D for example, will be written: ^3a, ^3b, ^3c, etc.

Axes, like sliding parts, can be designated by the capital letters of five or six different alphabets. Etruscan A, B, C; Roman A, B, C; writing $\mathscr{A}$, $\mathscr{B}$, $\mathscr{C}$, etc.

But once the alphabet has been decided, all pieces, be they fixed or moving, which are related to that axis, must be indicated by a capital letter of the alphabet given to the axis itself.

By this means, one can easily distinguish on the most complicated drawing, the axis to which any particular part belongs. In practice, the characters of the alphabet which are allotted to each axis should be selected in such a way that two axes on which the pieces cross each other should never be marked with the same character.

In the execution of the diagrams of the *analytical machine*, in a few cases only would it have been advisable to use four or even more different alphabets. In fact, no more than three were used.

The relative position of the pieces, whether fixed or moving, belonging to axes or to sliding pieces, can always be seen on each projection if the following rule is adopted.

Begin by lettering the most remote piece belonging to the axis, choosing a suitable letter; then, and in order of proximity, mark each other piece with a letter which does not come lower in the alphabet than any used previously.

Here, in brief, is the application of this notation system to the description of the Swedish machine.

In the series of diagrams now exhibited before the *Académie*, the only sections of the notation system used by Mr Henry P. Babbage are the *letters*, the *trains*, and the *cycles*.

The main diagram A, more than four metres in length, relates to the cycle. It contains about 700 vertical columns. The *time of motion* of each piece, and the *time of action* of each of its *working points*, can immediately be seen with the greatest of ease.

The red letters at the top of the columns indicate the connection of the *cycles* with the *trains*.

The *cycle*, or the period of the machine, has 96 time units. During the cycle, one calculation is completely executed and printed.

On diagram B all the several hundred moving pieces of the machine are clearly marked.

Each of these pieces displays its *working points*. The way in which each one is acted on and acts on other pieces can be seen.

The handle, to the left of the *trains*, is the source of movement. After being subdivided, and recombined in many different forms, this movement comes to a stop to the right of the trains through the agency of the collective result of the action of all the parts of the machine.

A series of steel punches, bearing the numerical characters laid out by the mechanism according to the law of the table which is to be calculated, are to be seen line by line above a leaden plate. This plate, after each operation, is pressed against the punches and thus receives, without any possibility of error, the impression of the figures which are the final result of the calculation.

APPENDIX

Facsimiles of plates

LIST OF PLATES

Thirteen plates of the Analytical Engine[a]

[a] These plates, reproduced from *Babbage's calculating engines* (London, 1889), are not dated. Some of the proofs of the plates in the Babbage papers, Waseda University, are dated as follows: 3 and 4, 10 May, 1837; 8–10, April, 1837; 12, 4 December, 1836. See 'Sketch of the Analytical Engine', this volume, pp. 91–2, for a description of the plates.

[b] Plates 2–7 should be considered as a group. The complete plate is given in Plate 6, and with lettering in Plate 7. Plates 2–5 show various parts of the complete plate.

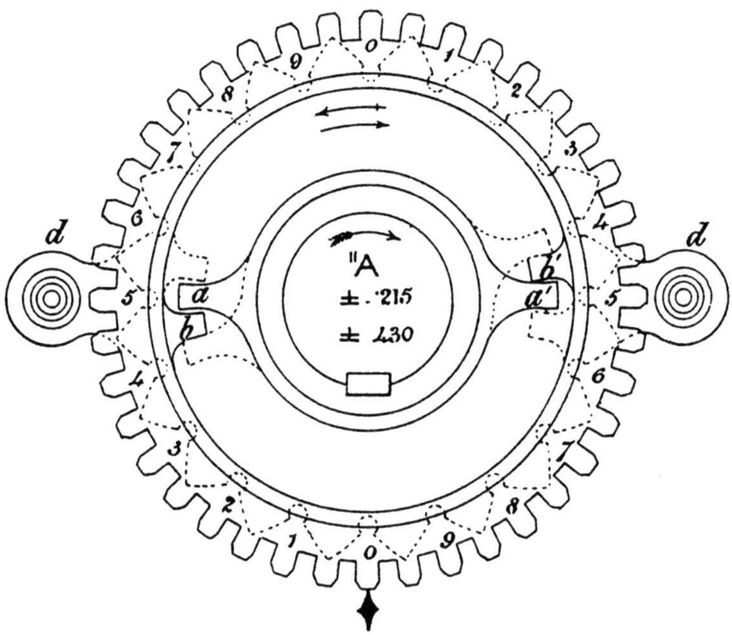

PLATE 1

Plan of the figure wheels for one method of adding numbers

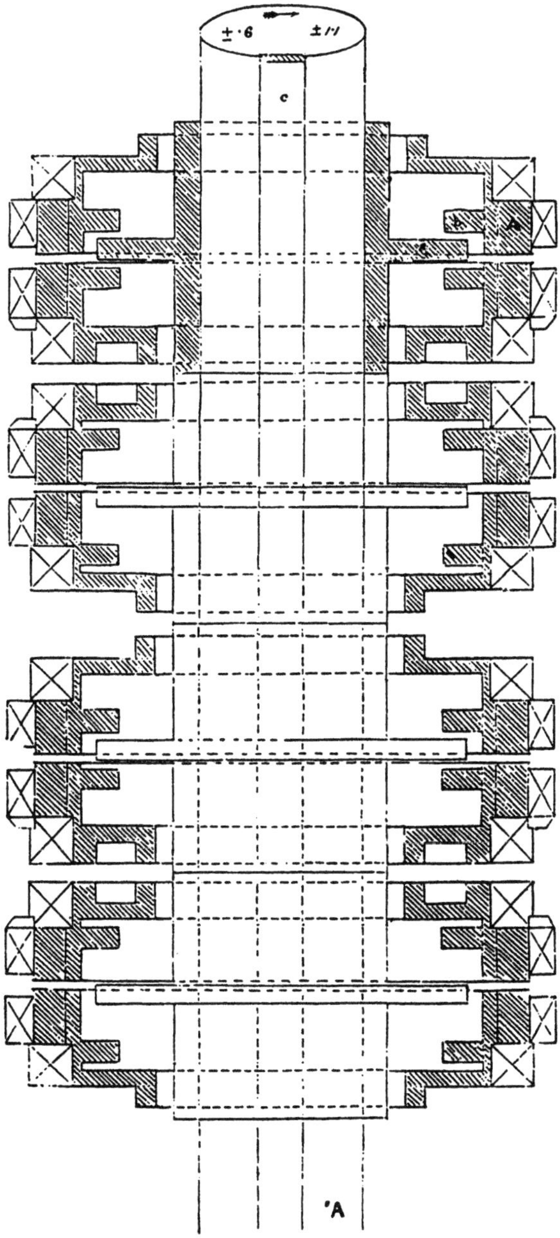

PLATE 2

Elevation of the wheels and axis for one method of adding numbers

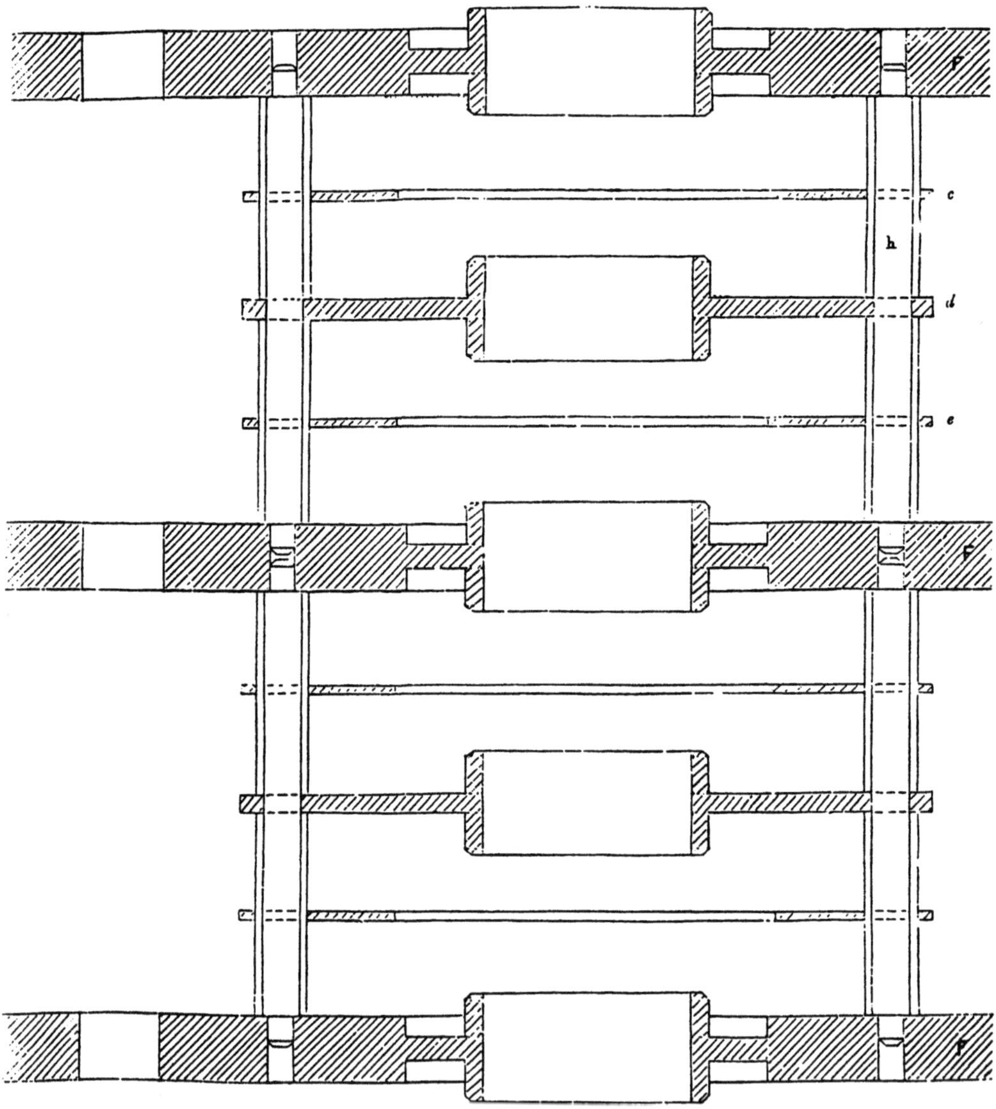

PLATE 3

Elevation of framing only for one method of adding numbers

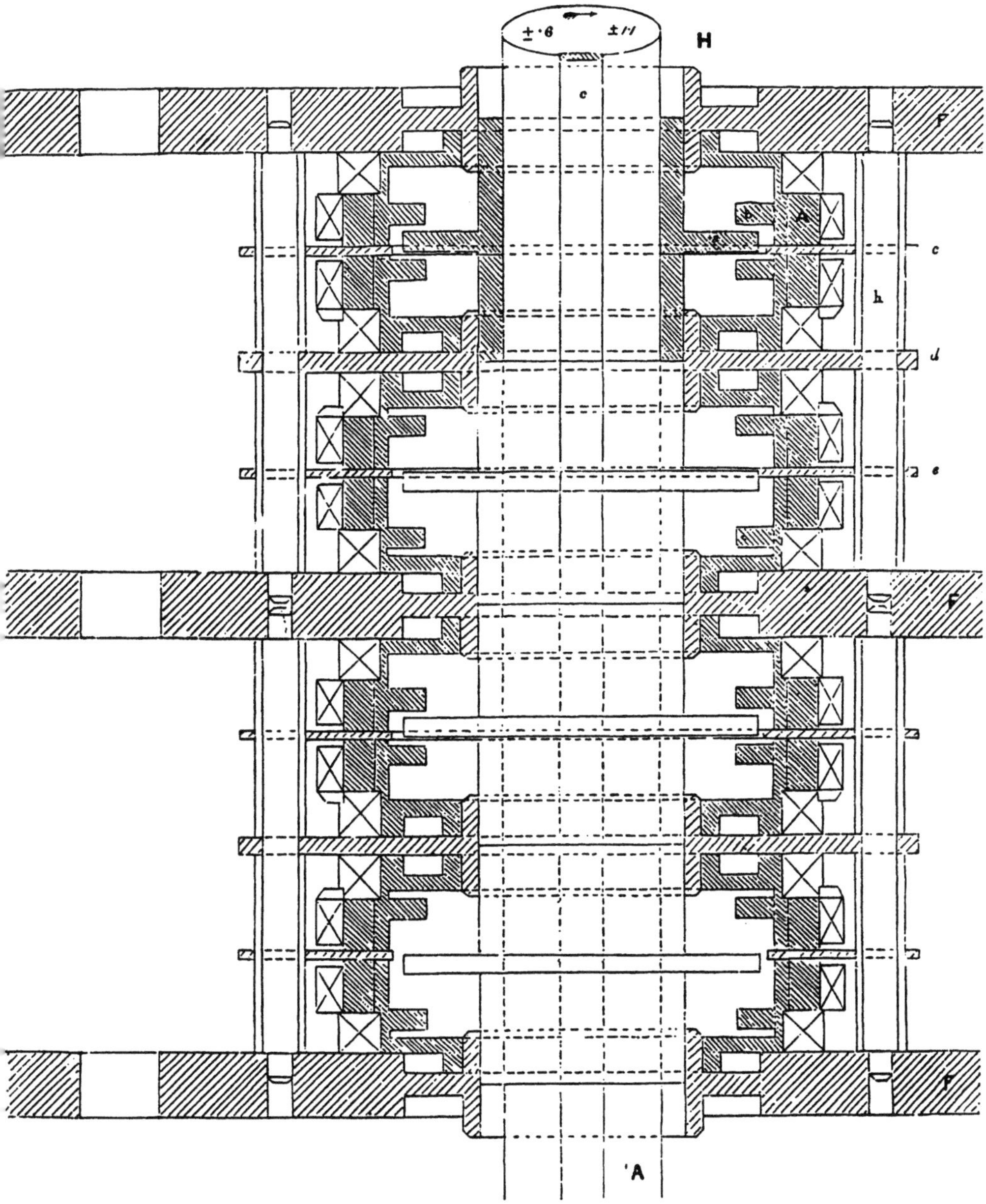

PLATE 4

Section of adding wheels and framing together

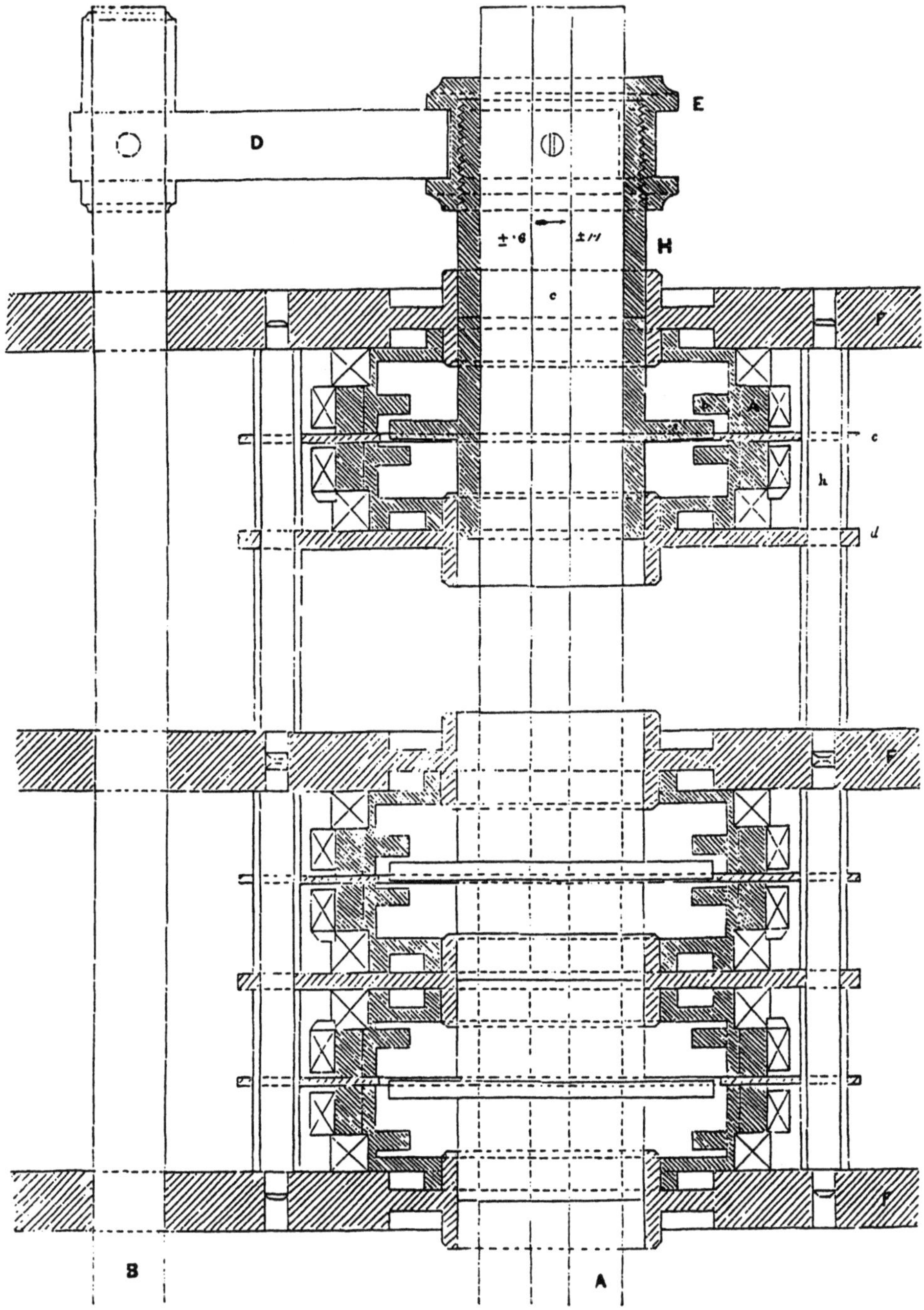

PLATE 5

Section of the adding wheels, sign wheels and framing complete

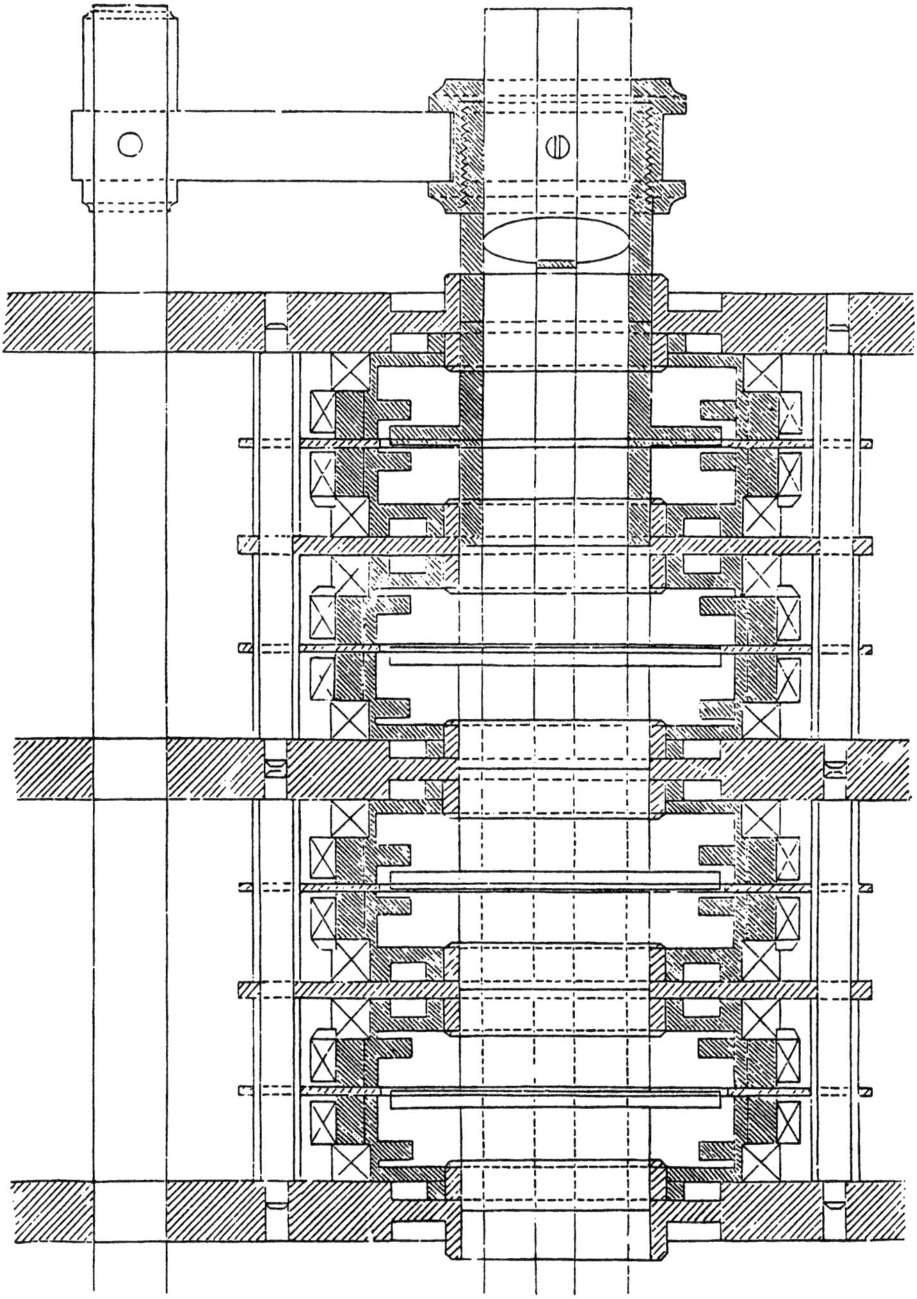

PLATE 6

Impression from the original wooden block

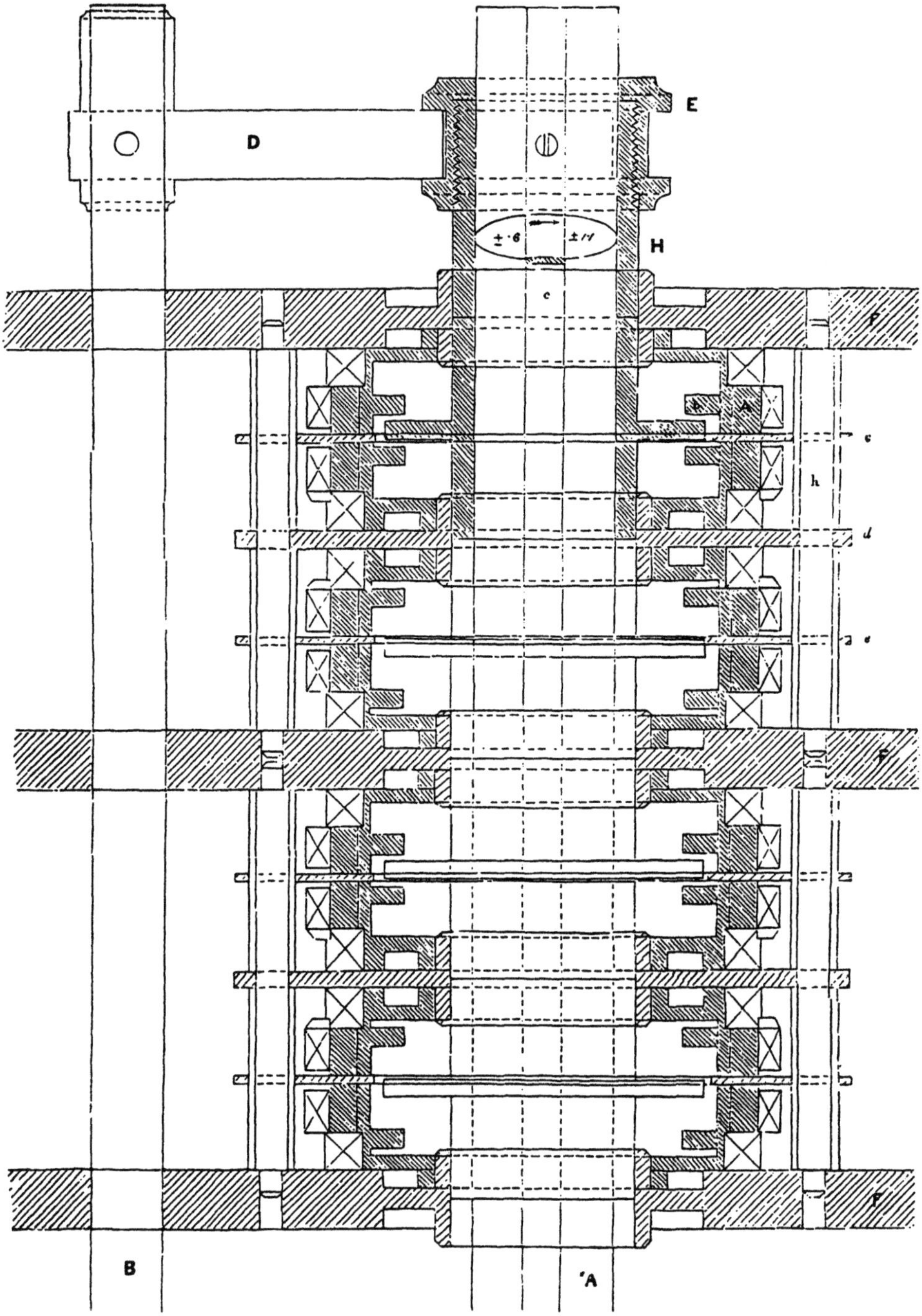

PLATE 7

Impression from a stereotype cast of No. 6, with the letters and signs inserted. Nos 2, 3, 4 and 5 were stereotypes taken from this

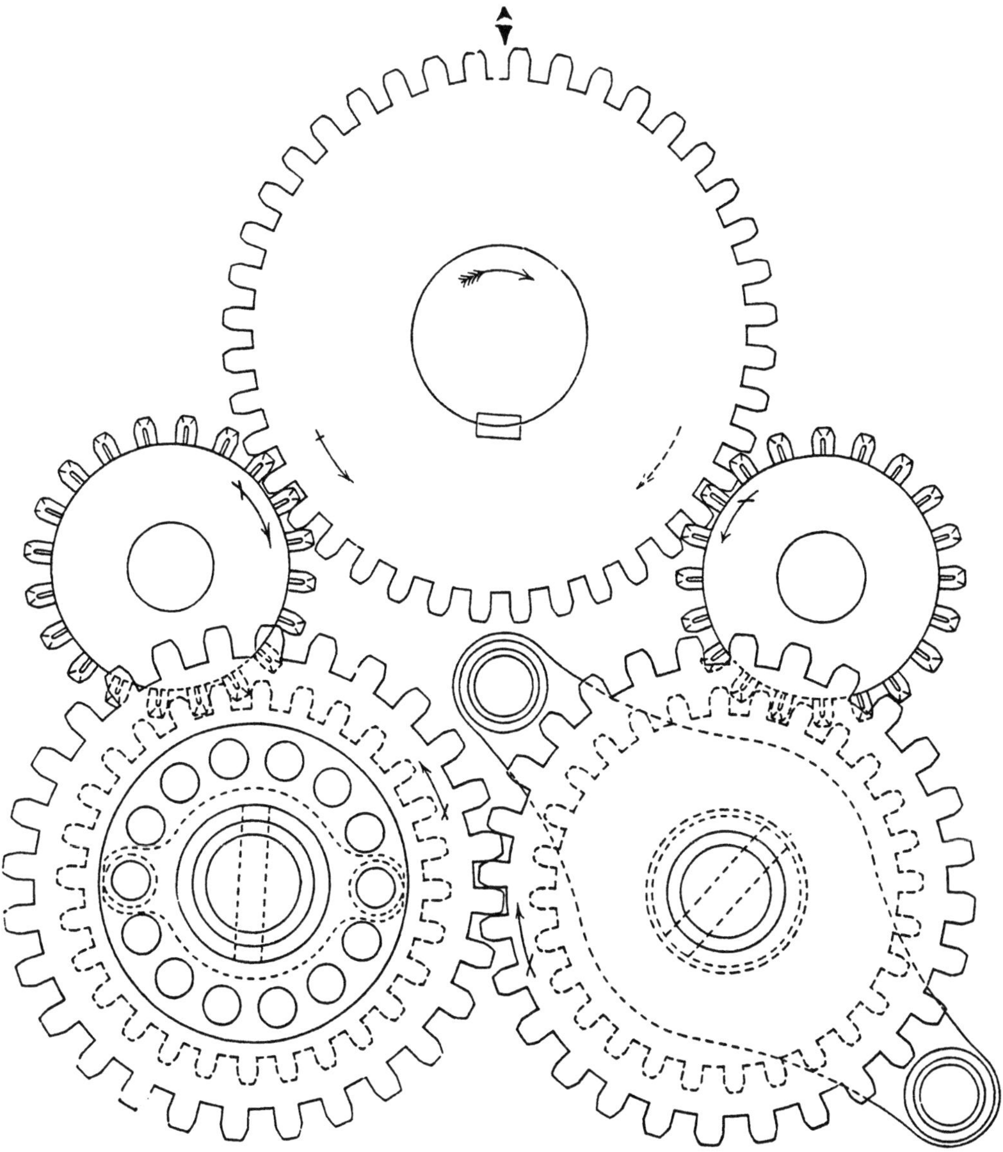

PLATE 8

Plan of adding wheels and of long and short pinions, by means of which *stepping* is accomplished

N.B. This process performs the operation of multiplying or dividing a number by any power of ten

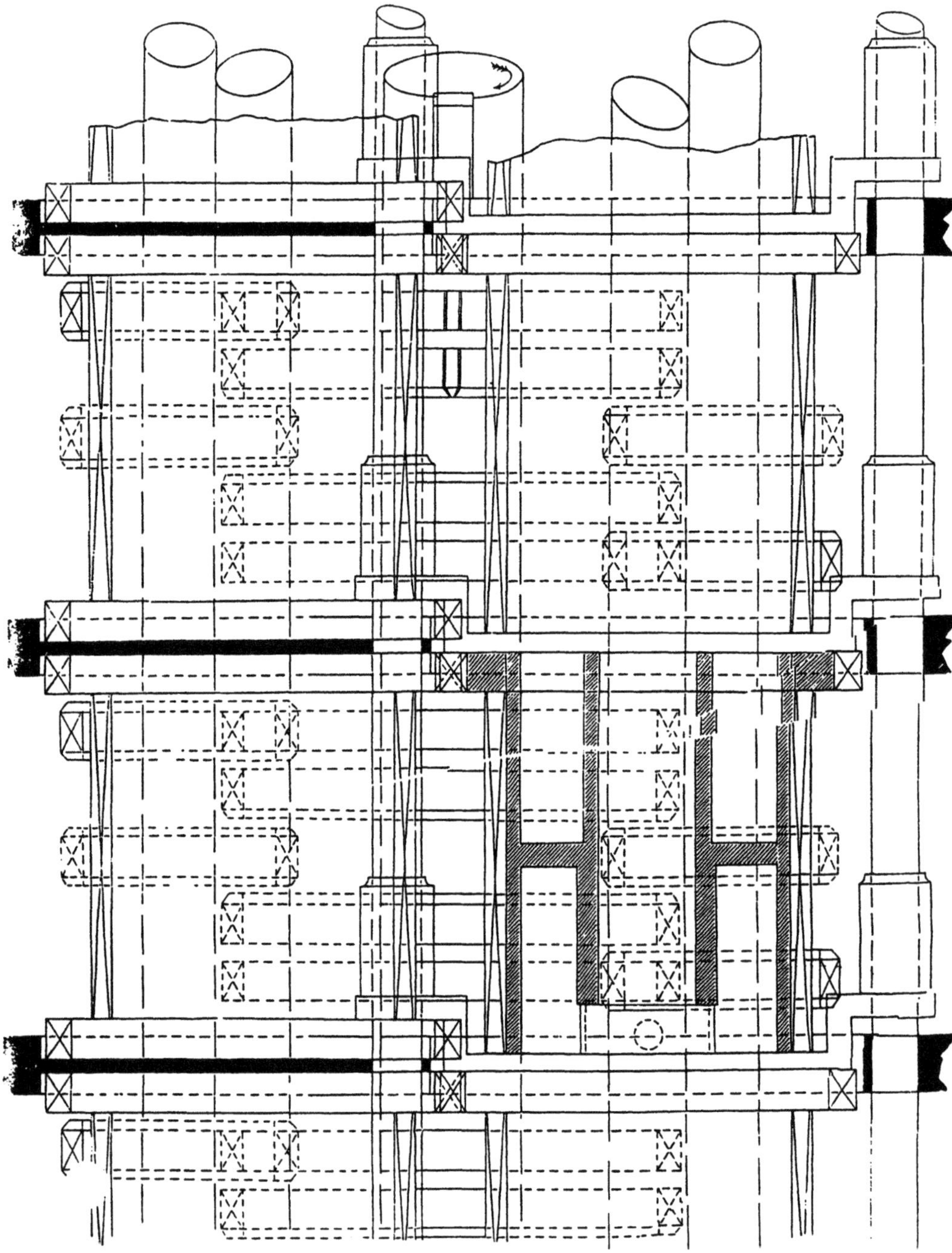

PLATE 9

Elevation of long pinions in the position for addition

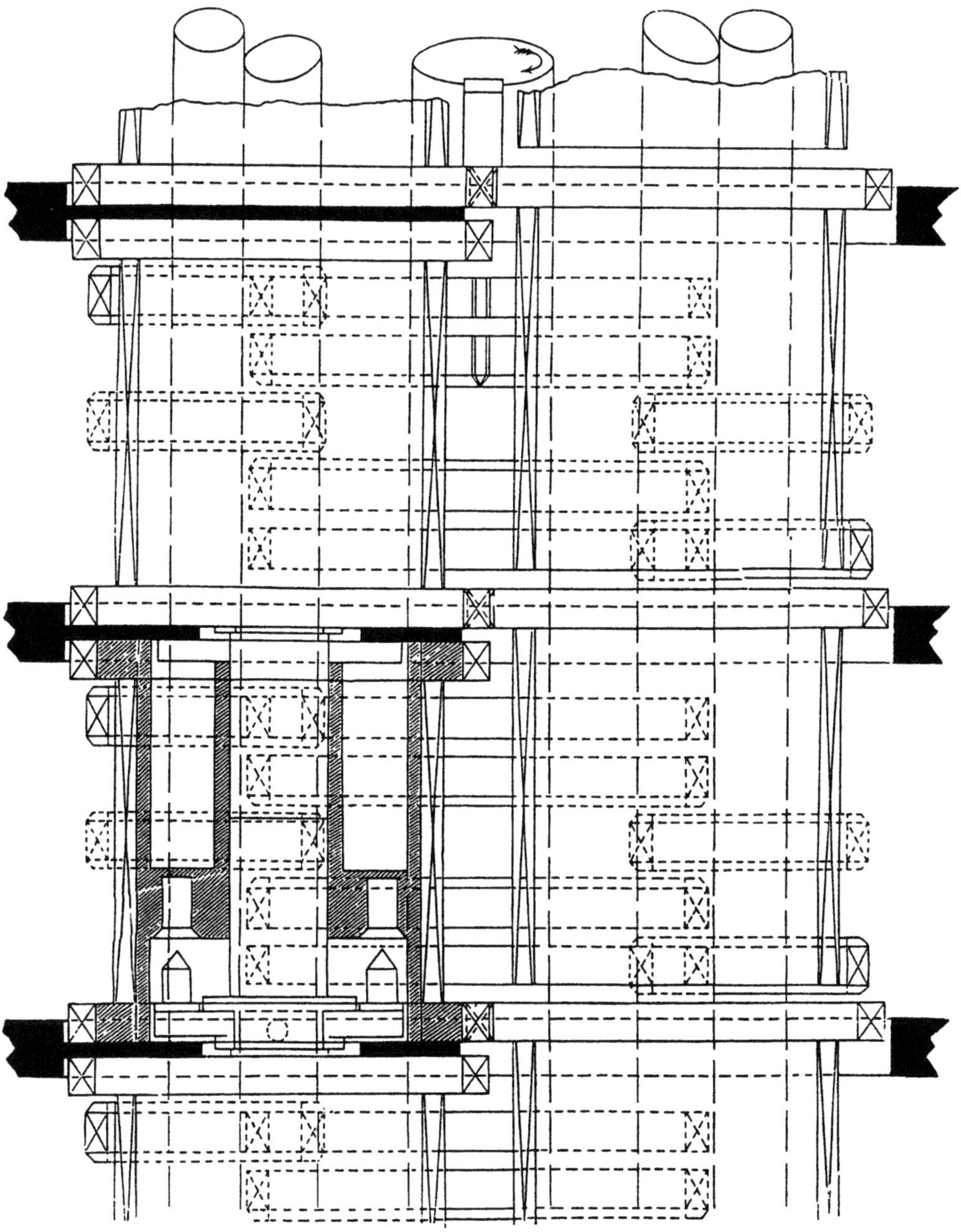

PLATE 10

Elevation of long pinions in the position for stepping

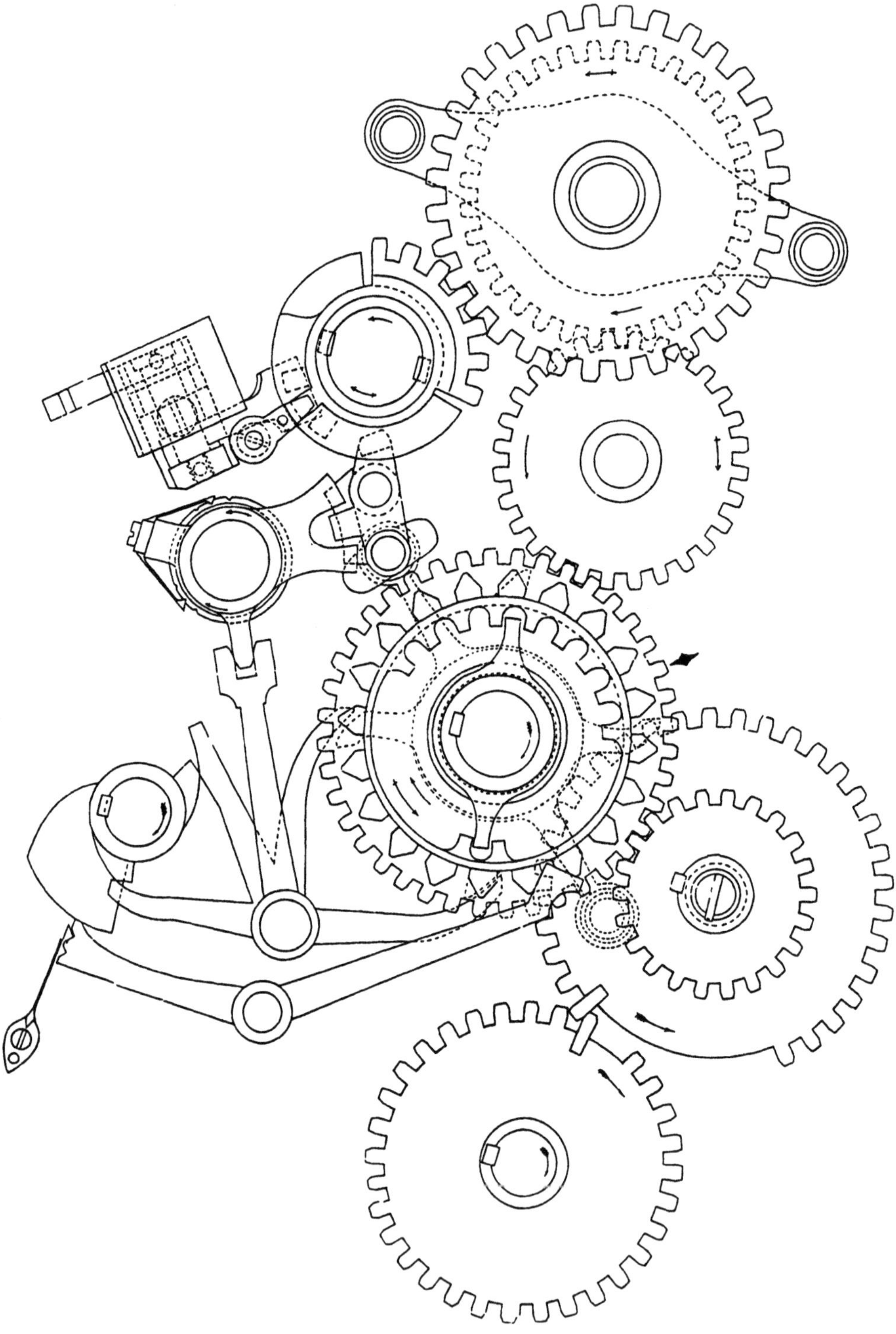

PLATE 11

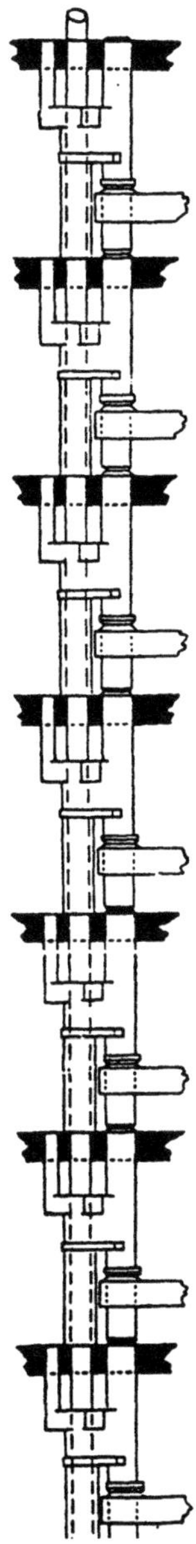

PLATE 12

Section of the chain of wires for anticipating carriage

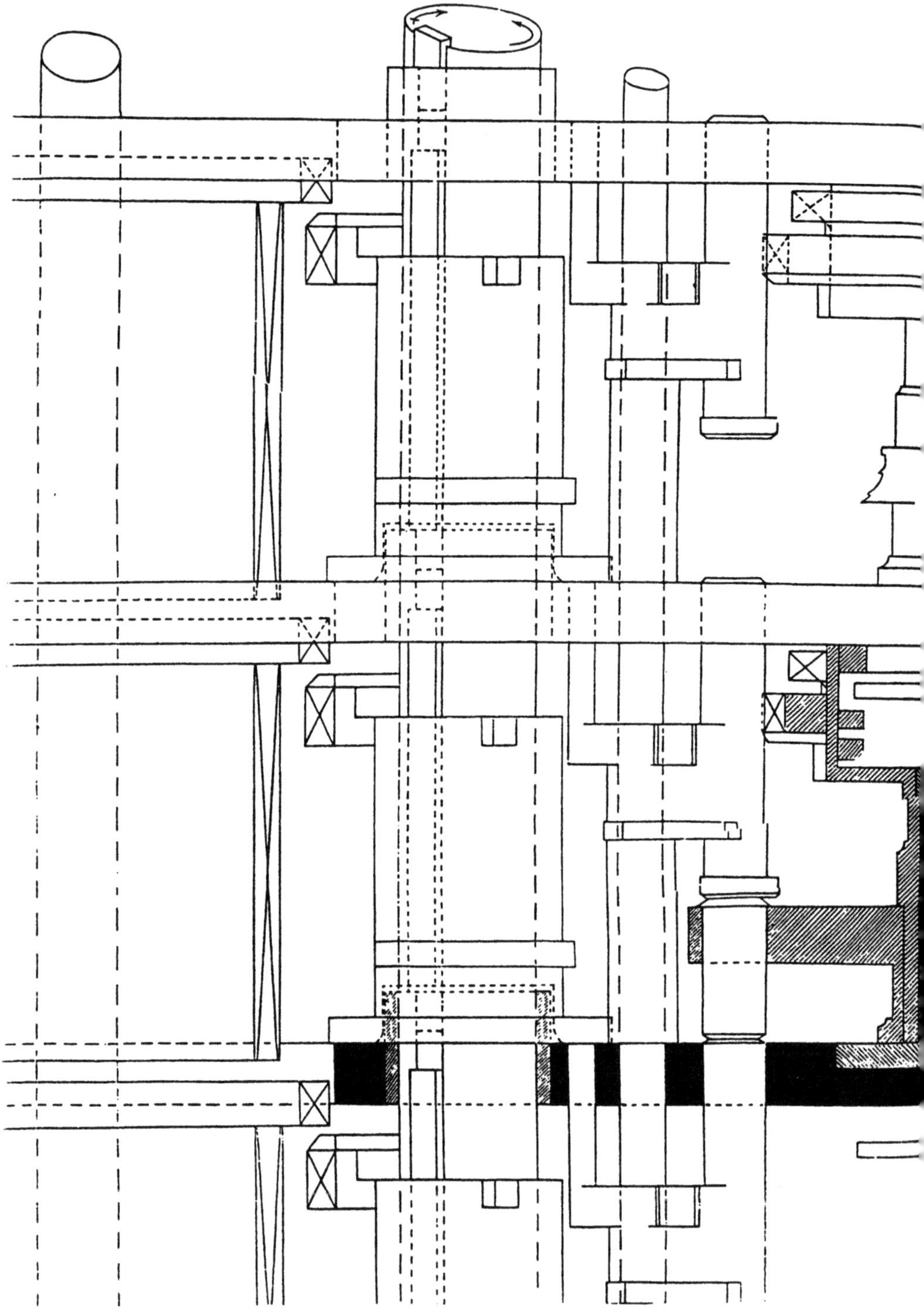

PLATE 13

Sections of the elevation of parts of the preceding carriage

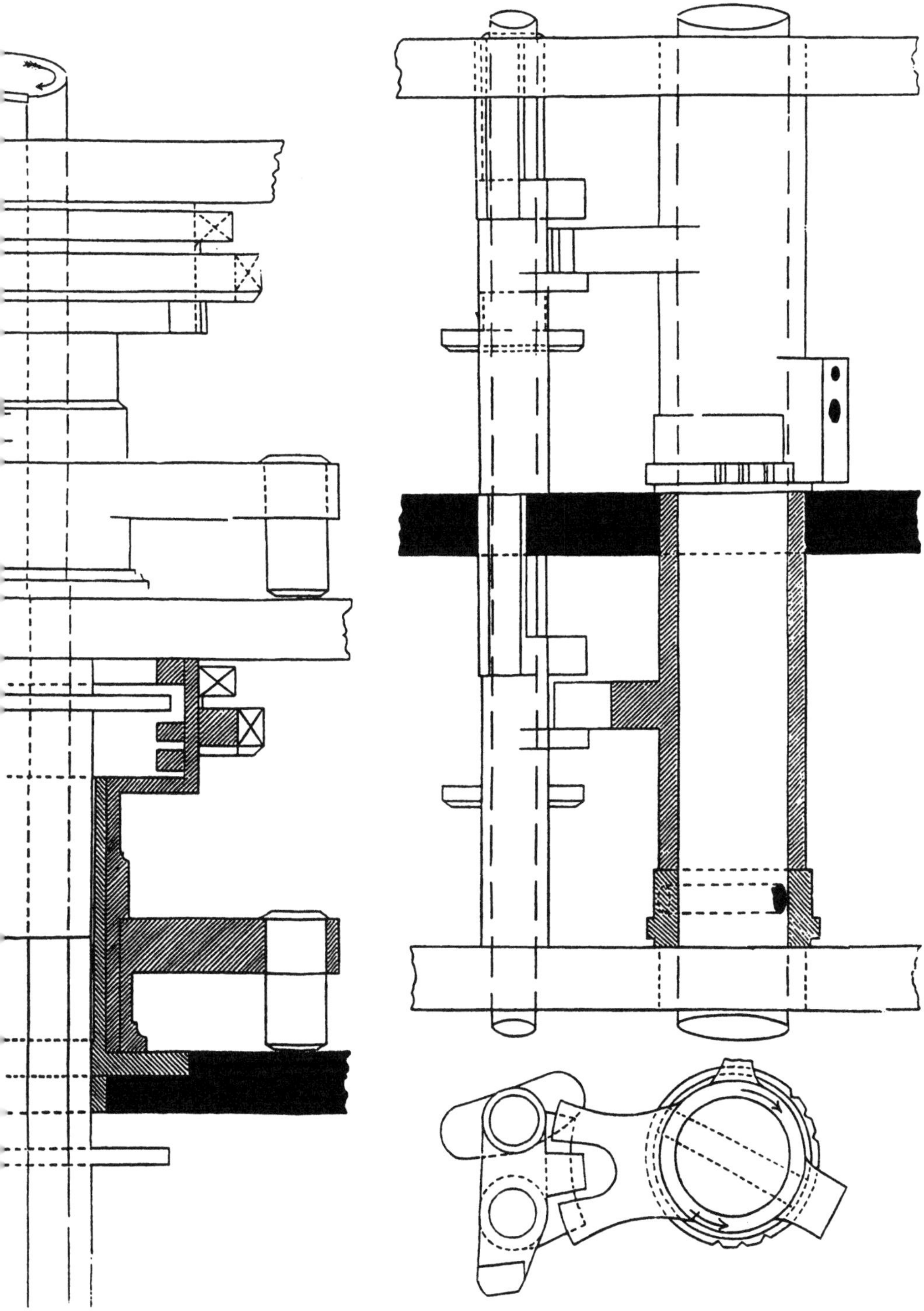

PLATE 13

Sections of the elevation of parts of the preceding carriage

General plan of M

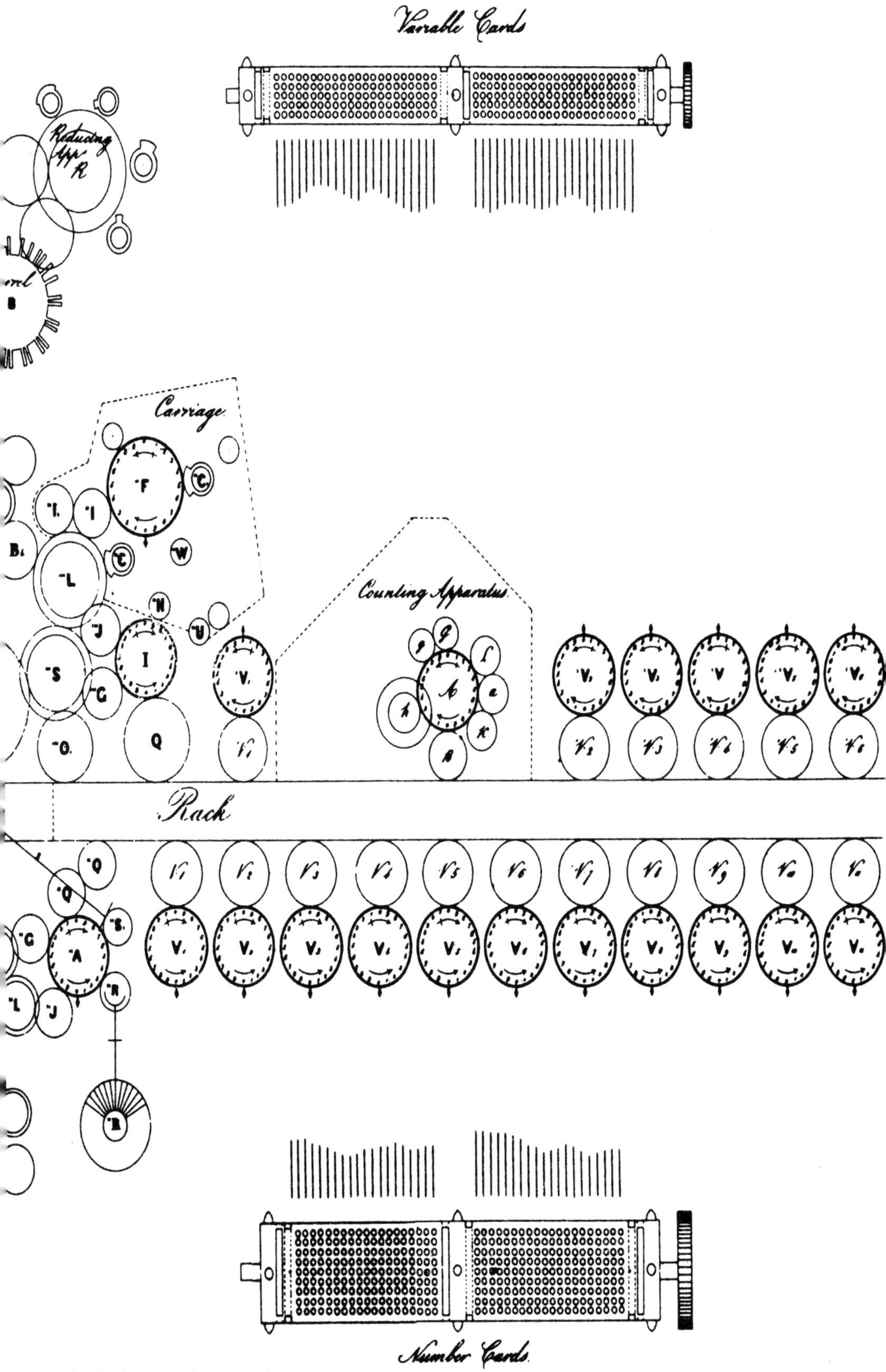

reat calculating engine, 1840.

For Product Safety Concerns and Information please contact our EU representative GPSR@taylorandfrancis.com Taylor & Francis Verlag GmbH, Kaufingerstraße 24, 80331 München, Germany

Batch number: 10725802

Printed by Printforce, the Netherlands